Webb Society Deep-Sky Observer's Handbook
Volume 5: Clusters of Galaxies

Webb Society Deep-Sky Observer's Handbook

Volume 5
Clusters of Galaxies

Compiled by the Webb Society
Edited by Kenneth Glyn Jones, F.R.A.S.
Written and Illustrated by George S. Whiston, Ph.D.

With a foreword by Professor George O. Abell
(Professor of Astronomy, University of California, Los Angeles)

Enslow Publishers
Hillside, New Jersey 07205

Library of Congress Cataloging in Publication Data

Main entry under title:

First published under title: The Webb Society
observers handbook.
Includes bibliographies.
Contents: v. 1. Double stars–v. 2. Planetary and
gaseous nebulae–[etc.] –v. 5. Clusters of galaxies.
1. Astronomy–Observers' manuals. I. Jones, Kenneth
Glyn. II. Webb Society.

QB64.W36 1979 523 78-31260
In the U.S.A. ISBN 0-89490-066-8 (v.5) AACR2

Manufactured in the United States of America

10 9 8 7 6 5

To Professor George O. Abell,
Deepest of Deep-Sky Observers,
With admiration and gratitude.

CONTENTS

CONTENTS (cont.)

LIST OF ILLUSTRATIONS

FOREWORD BY GEORGE O. ABELL

University of California, Los Angeles, September, 1981.

I am very honored to write a Foreword for the Webb Society's volume on clusters of galaxies. It is a subject close to my heart because I have been involved with the study of these galaxian archipelagos since the 1950s, when I began to identify them on the 48-inch Schmidt photographs taken for the National Geographic Society-Palomar Observatory Sky Survey.

What especially appealed to me about clusters of galaxies is that, at least for the very rich ones that comprise a more or less homogeneous sample, we can estimate relative distance from the faintness of their brightest member galaxies. If, as seems reasonable, the great clusters are tracers of the location of matter in general, their distribution gives us our first firm hold on the three-dimensional structure of the universe.

Indeed, in 1957 we could conclude that the cluster distribution showed large-scale isotropy and homogeneity -- consistent with the cosmological principle. On the small scale, on the other hand, the clusters are far from homogeneous in their distribution. Rather, clusters are clumped into larger aggregates of characteristic size 50 or 100 million parsecs (corresponding to values of the Hubble constant of 100 or 50 km s^{-1} Mpc^{-1}, respectively). At the time I called these systems second-order clusters; they are now known as superclusters. For more than a decade the reality of superclusters was a matter of much controversy; indeed, my conclusions were not, I think, accepted by the majority of astronomers. The relatively recent widespread acceptance of superclusters as real systems has been a vindication of some personal satisfaction to me.

Typical superclusters contain only about two individual clusters that could be classed as "great" -- for example, like those clusters in Coma Berenices or Corona Borealis. Some, including the Local Supercluster to which our own Local Group belongs, have not even a single great cluster (the Virgo cluster really would not qualify as a cluster rich enough for inclusion in my catalogue). A few superclusters, on the other hand, have more than a dozen great clusters. All have large numbers of lesser clusters and very many groups like the Local Group. They probably contain individual galaxies as well. I like to compare a supercluster to a great metropolitan area: perhaps one or a few major cities -- large centers of urban population -- surrounded by many smaller towns and suburbs, and generally some individual dwellings.

Significant but non-conclusive evidence (at this writing) suggests that there are no galaxies, groups, or clusters between superclusters; superclusters thus may well be the most fundamental concentrations of matter in the universe. Moreover, they appear to mark the end of the hierarchy. The uniformity of the distribution of galaxies and radio sources on scales much larger than 100 million parsecs suggests that higher order "super-duper" clusters do not exist.

Of course, there is far more to the study of clusters of galaxies than their distribution in space and their tendency to supercluster: their dynamical evolution and the likelihood of galaxies merging in the central regions of some clusters (galaxian cannibalism), the origin of SO galaxies, largely peculiar to rich clusters, the intracluster gas as revealed by X-rays and head-tail radio cources within clusters, and many other intriguing observations and theoretical analyses.

To me, however, the most exciting and profound unsolved problem is that of the origins of galaxies and clusters -- that is, which came first, galaxies, clusters, or superclusters? Dynamical studies in Great Britain and in the United States have shown that gravitating mass points starting with random initial positions and velocities in space eventually clump into systems resembling the groups and clusters collected together in superclusters that we observe. These theoretical models do not, however, account for all observed features in detail, especially the different populations of galaxy types in the rich regular clusters as opposed to the less organized irregular clusters, groups, and in the general field.

An alternative scenario is that of the Soviet cosmologists, Zel'Dovich, Doroshkevich, and others, who have suggested that gas condensations in the early universe formed superclusters first, and that clusters and subsequently individual galaxies were secondary and tertiary condensations. This model predicts that superclusters should be flat, like "pancakes", and indeed, there is much evidence, not by any means conclusive, that superclusters may be flat pancake-like systems.

The truth, of course, may be far more subtle and complex than any theoretical models considered to date. But what grander areas are there for research? Probably what we learn of the properties of clusters of galaxies over the coming years will provide vital links in the resolution of the questions of origins.

In a way it is too bad that clusters of galaxies are such disappointing visual telescopic objects. Only large-aperture, high speed optical systems reveal even the nearest, most conspicuous clusters as interesting to the novice. Rather few professional telescopes, in fact, produce pleasing photographs of clusters. We are looking at widely scattered objects of very low surface brightness. At cosmologically interesting distances, nothing (save stellar appearing quasars) is apparent to visual inspection through even the world's largest telescopes.

Yet, the elusive rich clusters of galaxies are the grand systems that probably carry the clues to the basic structure of the universe, as well as representing many exciting problems at the forefront of modern astrophysical research. It may take considerable effort to observe them, and their telescopic appearance may never impress your neighbors. Still, it is rewarding to find them when you consider their impact on our understanding of the universe as a whole.

Good hunting to you all, and I hope you enjoy and appreciate your glimpses of the cosmos as much as I do! With thanks.

General Preface

Named after the Rev. T.W. Webb (1807-1885), an eminent amateur astronomer and author of the classic Celestial Objects for Common Telescopes, the Webb Society exists to encourage the study of double stars and deep-sky objects. It has members in almost every country where amateur astronomy flourishes. It has a number of sections, each under a director with wide experience in the particular field, the main ones being double stars, nebulae and clusters, minor planets, supernova watch and astrophotography. Publications include a Quarterly Journal containing articles and special features, book reviews and section reports that cover the society's activities. Membership is open to anyone whose interests are compatible. Application forms and answers to queries are available from E. G. Moore, Publications Officer Webb Society, 1, Hillside Villas, Station Road, Pluckley, Kent. TN27 0QX England.

Webb's Celestial Objects for Common Telescopes, first published in 1859, must have been among the most popular books of its kind ever written. Running through six editions by 1917, it is still in print although the text is of more historical than practical interest to the amateur of today. Not only has knowledge of the universe been transformed by modern developments, but the present generation of amateur astronomers has telescopes and other equipment that even the professional of Webb's day would have envied.

The aim of the new Webb Society Deep-Sky Observer's Handbook is to provide a series of observer's manuals that do justice to the equipment that is available today and to cover fields that have not been adequately covered by other organisations of amateurs. We have tried to make these guides the best of their kind: they are written by experts, some of them professional astronomers, who have had considerable practical experience with the pleasures and problems of the amateur astronomer. The manuals can be used profitably by beginners, who will find much to stimulate their enthusiasm and imagination. However, they are designed primarily for the more experienced amateur who seeks greater scope for the exercise of his skills.

Each handbook is complete with regard to its subject. The reader is given an adequate historical and theoretical basis for a modern understanding of the physical role of the objects covered in the wider context of the universe. He is provided with a thorough exposition of observing methods, including the construction and operation of ancillary equipment such as micrometers and simple spectroscopes. Each volume contains a detailed and comprehensive catalogue of objects for the amateur to locate and to observe with an eye made more perceptive by the knowledge he has gained.

We hope that these volumes will enable the reader to extend his abilities, to exploit his telescope to its limit, and to tackle the challenging difficulties of new fields of observation with confidence of success.

Preface
Volume 5: Clusters of Galaxies

With the publication of this volume the Webb Society concludes its declared intention of surveying the heavens through the eyes of the modern amateur astronomer. With the enormous amount of information made available to him by his professional colleagues, and with the benefit of present day technology, the amateur of today can attempt feats of visual observation which would have severely taxed the skill and resources of some of the leading professional observers even at the beginning of the present century.

The recognition that 'island universes' really exist beyond the confines of our own Galaxy was not finally established until as late as 1924 but since then, we have found that extra-galactic systems proliferate in numbers that are virtually uncountable, and that they extend as far as light itself can travel. However, it is a comparatively recent conception which recognizes that, remote and immense as the galaxies are, they exist, not as single 'islands', but as members of gravitationally bound super systems of varying size and population - in 'clusters of galaxies'. And it is more and more apparent, too, that it is these cluster systems, and not the individual galaxies, which are the fundamental building blocks of the universe.

To observe this elemental structure in its actuality requires the highest refinement of the observer's craft, and perhaps we can claim that with this publication we arrive at the culmination to which all previous volumes in this series have been striving. There is plenty of scope for the amateur here: the Virgo Cluster can be explored with profit even in a 4-inch reflector, and this magnificent region is described in generous detail in the following pages. One has to admit, of course, that in this particular aspect of deep-sky observing it will be the larger apertures which will produce the higher dividends. Very many amateurs, by the time they have served their observing apprenticeship, will have telescopes of more than adequate performance: by competent application, and with the aid of this Handbook, they can anticipate prospects of exotic interest and rare delight.

In Part One the theoretical background to the subject is given in some depth since this is a field which heretofore has generally been held to be beyond the scope of amateur activities. The text, we trust, is clear, concise and adequate to its purpose and is supplemented by extensive references for further reading in the Bibliography. A chapter on Visual Observation covers the special techniques required in this taxing field which requires not only the highest skill but even, on occasions, as the author puts it, "grim determination", to achieve success.

In Part Two the three detailed catalogues of the Virgo Cluster, Abell Clusters and Groups of Galaxies provide what we claim to be an unique collection of celestial objects for amateur observation.

The Catalogues are supplemented by numerous finding charts for identification and more than 120 field drawings made by Webb Society members with telescopes ranging from 5 to 36 inches in aperture.

Here then, is a feast of expert factual data, and a display of visual celestial beauty to whet the appetite of the most fastidious amateur astronomer. Readers have only to seize the opportunity and the deepest regions of the heavens are theirs to survey.

The text of this book has been written entirely by Dr G.S.Whiston, who has also executed the many line-drawings and prepared the telescope field drawings for reproduction. George Whiston is a mathematician who has published more than 20 papers on various aspects of pure and applied mathematics. He is currently working on some of the problems of structural dynamics in nuclear power reactors in the U.K. but declares that he finds astrophysics and cosmology rather more rewarding to study. He is also an active deep-sky observer whose work with a largely home-built reflector is amply exhibited in these pages.

The Editor - and the Society - are deeply grateful to Dr Whiston for his work in compiling this volume. Thanks are also due to E.S.Barker, F.R.A.S., and E.G.Moore, F.R.A.S., for their valuable role in proof-reading, and to Ridley M. Enslow, President of Enslow Publishers, who has done so much to ensure the successful publication of all volumes of this Handbook.

Finally, we wish to express our deepest gratitude to Prof. G.O.Abell for honouring us with the acceptance of the dedication of this volume, and for so generously contributing a Foreword. Professor Abell needs no introduction in any field of astronomy; in its widest realms - among the Clusters of Galaxies - his is a household name, and we are grateful and delighted that he has given our modest efforts such eminent support.

Author's Acknowledgements.

The author is grateful to Dr Jon Godwin of the University of Oxford for supplying film copies of galaxy cluster fields taken with the 48-inch Palomar Schmidt: these were basic to the preparation of this volume. Thanks are also due to Malcolm Thomson for compiling some of the identification tables (and for his astronomical correspondence over the years), to Ron Buta for his superb McDonald Observatory visual observations, and to Bob Argyle of RGO for precessing positions in the list of Abell clusters. Ken Glyn Jones and Ed Barker are also due thanks for their guidance and encouragement during the production process. Last, but not least, thanks to the typist, Miss Tina Jones for her patient work.

PART ONE : CLUSTERS OF GALAXIES

1. INTRODUCTION

Previous volumes in this series of handbooks dealt with the visual observation of astronomical objects of increasing size and distance. Volume 1 covered the measurement of double stars and Volume 2 dealt with the observation of planetary and gaseous nebulae. Such objects lie relatively nearby on the astronomical scale, being constituents of our Milky Way Galaxy, and lying at distances within the Galaxy limited by the obscuration caused by intersteller dust. Together with the open star clusters discussed in Volume 3, one can say that the obscuration limits their distances to, say, at most 10 Kpc or about 30,000 light years. Of course, a few systems of the above type might be observable at larger distances if, for example, they lie above the plane of the Galactic disc or are seen through 'holes' in the intersteller dust. Volume 3 also dealt with the observation of globular clusters which, although belonging to the Milky Way Galaxy, lie outside of the main system, forming part of the spherical halo of the Galaxy which has a diameter roughly equal to the the disc component, about 30 Kpc or 100,000 light years. In Volume 4, covering the visual observation of isolated or 'field' galaxies, the distances involved began to become unimaginably large, ranging from about 0.7 Mpc to the nearest large galaxy to the Milky Way - M.31, the Great Andromeda Nebula - to about 100 Mpc, the average distance of the faintest galaxies that are visually detectable in relatively large amateur telescopes and shine at about the 15th magnitude - about a million times fainter than the brightest naked eye stars.

At a scale of more than a few tens of megaparsecs, individual galaxies lose their identities, much as do the molecules in a gas for everyday physics. On these scales, the collective dynamics of galaxies is fashioned by universal gravitation which tends to gather galaxies into relative concentrations over a galactic field representing an imaginary initially uniform distribution throughout space. Such relative concentrations are called clusters of galaxies and these physically bound objects probably represent the ultimate scale of physical objects that exist in the universe. Rather than take the view that clusters of galaxies formed from an initially uniform distribution of galaxies, we shall adopt the viewpoint that clusters represent the fundamental blocks of matter. It is possible that clusters themselves are clustered, and we shall discuss this second order clustering of galaxies in a later chapter. However, according to present knowledge, higher order hierarchies of clusters probably do not exist in nature. The second order clusters, or superclusters, that have been observed do not entirely dwarf their constituent clusters in scale and it may be that the constituent clusters represent mere condensations in large clusters. For example, a supercluster may only contain at most 3 or 4 constituent clusters and have a diameter of the order of 40 or 50 Mpc, whilst rich clusters have diameters of the order of up to 10 Mpc. According to current cosmological thinking, it is possible that field galaxies do not exist in nature. Only objects with masses of large clusters of galaxies seem to be able to exist through the early radiation dominated era of the universe in the Hot Big-Bang models. Indeed, observations of the distribution of extremely distant galaxies which swarm in their thousands over very deep large-aperture photographic plates have been shown to be entirely consistent with universal clustering. That is, very distant and intermediate clusters of galaxies may overlap on the celestial sphere to such an extent that the probability that a 'field' galaxy chosen at random may belong to a nearby obvious concentration of galaxies of a similar appearance is relatively high.

Clusters of galaxies exist with populations ranging from 2 galaxies to about 10,000 galaxies. It may be that the poor clusters represent sub-condensations in larger more populous clusters. Nearby clusters have very large apparent size, and may not readily be recognised as such without reliable distance estimates. For example, the Milky Way Galaxy is part of a cluster of about two dozen galaxies (the Local Group) with two large systems - the Milky Way and M31, and a few smaller galaxies such as M33, the companions of M31-M32 and NGC.205, and the Magellanic Clouds, the rest being dwarf systems. A similar but more distant cluster is dominated by M81. The nearest giant cluster lies beyond the constellation of Virgo, and contains thousands of galaxies. Because of its proximity it is of large apparent size, and it can be traced beyond the constellations of Coma Bernices, Canes Venatici and Ursa Major. The cluster contains many condensations or subgroups, and it is thought that our Local Group is but an outlying condensation towards the edge of the system, the latter conclusion following from the fact that there is no corresponding cloud of galaxies in the galactic antidirection.

In this volume, we take the largest step outward in space and time that is possible for the amateur astronomer to make by discussing the visual observation of groups (poor clusters) and rich clusters of galaxies. Our distance scale will range from the nearest rich cluster of galaxies, the Virgo Cluster, which lies at a distance of about 20 Mpc (if H_o = 50 km sec^{-1} Mpc^{-1}), to the Ursa Major I cluster (Abell 1377) which has a distance of 306 Mpc - about 1,000,000,000 light years and the Corona Borealis cluster (Abell 2065) which lies at the immense distance of 1,500,000,000 light years. The distances quoted above are, of course, redshift distances, corresponding to mean redshifts of 0.003 for the Virgo Cluster and 0.051 and 0.07 for A1377 and A2065. It is interesting to compare them with the redshifts of the most distant clusters of galaxies yet measured with the 200" Hale reflector on Mt Palomar, which have redshifts of about 0.9 (when the simple Hubble distance redshift relation is strictly inapplicable). Thus the modestly equipped amateur astronomer does not lag ridiculously far behind the professional in his ability to detect normal galaxies at vast distances. However, it should be noted that if the redshifts of quasars are cosmological in origin and not due to some new law of nature associated with the incredible energies expected to be found in their hearts, their distances will dwarf those of all observable normal galaxies. The largest quasar redshifts are of the order of 3.5. Using redshift as a cosmological evolution parameter, it is thought that the universe of galaxies and clusters of galaxies crystallised at a redshift of about 5.0. These redshifted quasars are more than 'half as old as time' and lie at distances of tens of thousands of megapasecs.

The clustering of galaxies is obvious on cursory inspection of the most elementary star atlas. For example, limiting oneself to galaxies in Messier's catalogue, their concentration in an area of the sky bounded roughly by lines through Denebola (beta-Leonis), Spica (alpha-Virginis) and the 6th magnitude star rho-Virgonis is obvious. Indeed, this region is labelled 'The Realm of the Galaxies' in the monthly star charts published in '<u>Sky and Telescope</u>'. A map with more detail is to be found in Norton's Star Atlas*. In a more detailed atlas, such as '<u>Atlas Coeli</u>',** which depicts most of the galaxies brighter than the 13th magnitude, the concentration of galaxies in this region is very marked, and the high density can be traced for about 30° in a band centred on the Realm of the Galaxies stretching from southern Virgo to Ursa Major. The distribution is dominated by the Virgo Cluster, and the cloud contains galaxies with generally decreasing distances as one progresses northwards. This huge cloud is known as the Local Supercluster, and appears to consist of intermeshed groups and clusters distributed roughly in a disc shaped swarm, with our Local Group situated towards the edge. As progressively fainter galaxies are included in the census of the cloud, its richness becomes very high indeed. For example, all NGC and IC objects are plotted in the SAO*** star atlas, and in this census which includes galaxies as faint as magnitude 15.5, the crowding together of the nebulae is so pronounced that identification from, say, the RNGC**** becomes difficult. The SAO atlas is an almost indispensable tool to the amateur deep sky observer because of its high stellar magnitude limit and the above mentioned

*G.E. Satterthwaite, P. Moore and R.G. Inglis (eds.) 'Norton's Star Atlas and Reference Handbook,' Cambridge, Mass., Sky Publishing Corp.

**Becvar, A. 'Atlas of the Heavens--II, Catalog 1950.0 (Atlas Coeli Skalnate Pleso--II)'. (see References and Bibliography.)

***'Smithsonian Astrophysical Observatory Star Atlas of Reference Stars and Non-Stellar Objects.' (see References and Bibliography.)

****J.L.E. Dreyer 'New General Catalogue of Nebulae and Clusters of Stars.' (see References and Bibliography.)

plotting of all NGC and IC objects. If one browses through its pages, the clustering of galaxies is everywhere apparent, consisting of knots of one or two galaxies, through richer concentrations such as the NGC.3158 group and the NGC.5416 group, to incredibly rich clusters such as the Coma Bernices Cluster, Abell 1656, and the Virgo Cluster, where the galaxy symbols are so crowded as to overlap into a dense swarm.

Because of the possible physical association of some clusters as subcondensations in large clusters and the large apparent size of nearby poor clusters, this volume will be concerned with a sub-class of all clusters of galaxies that are theoretically available to amateur telescopes. We shall further narrow our field of interest to include (a) those Abell clusters of galaxies available to northern hemisphere amateur observers, (b) the central region of the Virgo Cluster and (c) those poor clusters (groups) of galaxies which contain at least 3 or 4 galaxies in a field of at most $\frac{1}{2}^{\circ}$. In the latter subset, we can, of course, only consider a representative sample.

As a collection of (mostly) galaxies, the NGC is a non-homogeneous, incomplete survey of galaxies down to about the 15th magnitude, being just about complete for galaxies brighter than the 13th magnitude. The catalogue is mainly composed of nebulae and clusters found visually by the Herschels. By the very method of its compilation, it surveys some regions of the sky more efficiently than others, and to fainter magnitude in some regions than others. Such incompleteness is a natural consequence of the extreme difficulty of the mammoth task undertaken by these great observers in their survey of the entire sky, many objects being at the limits of visibility. In the next chapter, we shall briefly review more homogeneous surveys of the distribution of galaxies upon the sky, to increasingly faint magnitude limits. Such surveys form the basis for the current knowledge of the distribution of matter in our cosmological neighbourhood of space and will hopefully serve as a map for the explorations to be embarked upon in the catalogue sections of this volume. As the survey magnitude limits increase, one penetrates deeper into space according to the local inverse square law of light-attenuation with distance which can be expressed by the formula:-

$$\log_{10}(d_o/d) = -0.2(m - m_o)$$

A given sample of galaxies to apparent magnitude 'm' will, in general, contain galaxies of varying distances; from nearby dwarf galaxies to extremely distant supergiants, but given a large enough sample, the distance of a randomly selected galaxy will increase as the apparent magnitude increases in accordance with the above formula. If galaxies are distributed uniformly in space, that is, with a constant number of galaxies per cubic megaparsec, the number of galaxies brighter than apparent magnitude 'm' within a cone, with base area one square degree on the sky should increase according to the simple law:-

$$\log_{10}(N(m)) = 0.6(m - m_1)$$

Here m_1 is the apparent magnitude such that $N(m_1) = 1$, called the 'space density parameter'. That is, one has to penetrate to a depth given by the first formula with $m = m_1$ until one can expect to find a galaxy within the cone. It turns out that m_1 is about 15. Therefore one has to penetrate to a distance of about 100 Mpc until one can guarantee that a galaxy will be contained within the solid angle subtended by the cone. Extragalactic space is therefore populated by about 1 galaxy per hundred cubic megaparsecs for $m_1 = 15.0$. The slope of a plot of $\log(N(m))$ against m should therefore be 0.6 if space is uniformly populated by galaxies, any deviation of an observational plot from a straight line with this slope is a measure of galaxy clustering. It turns out that for small values of m, surveys of nearby extragalactic space, the distributions are far from uniform, an indication of the heavy clustering of galaxies at small cosmological scales. However, as the magnitude limit increases, say to m = 18.0, the distribution of galaxies becomes more closely uniform and isotropic, that is, independent of direction. This justifies the usual assumptions of the homogeneity and isotropy of the large scale distribution of matter in the universe.

In the next chapter, the various progressively deeper surveys of the distribution of galaxies in space, will be discussed with regard to the mapping of our cosmic neighbourhood. The chapter is mainly concerned however, with a description of G.O. Abell's great survey of the distribution of rich clusters of galaxies in space, his catalogue of clusters of galaxies*, and his conclusions regarding the universal distribution of clusters and the tendency of clusters to congregate in superclusters. More recent work on superclustering is also discussed, since many of the associations contain galaxy clusters forming part of our catalogue of clusters, and the discussion therefore places these visually fascinating objects in their true context.

The following chapter is a brief survey of the physical nature of clusters of galaxies. Emphasis is placed upon the physics of rich clusters since many of the catalogue objects can be so classed. The chapter starts with a review of some theories of cluster formation in a cosmological context, and then briefly discusses the physics of individual rich clusters.

*G.O. Abell 'The Distribution of Rich Clusters of Galaxies'. (see References and Bibliography.)

Chapter 4 is concerned with the visual observation of clusters of galaxies. The emphasis here is on how to try to push one's observing technique to the ultimate pitch in attempting to observe as deep into space as one's equipment will allow. The main difference between field galaxy and galaxy cluster observation is in identification. It is only too easy to pick up galaxies within easy reach of one's telescope in a cluster, but as observers of the Virgo Cluster (where a random sweep yields many galaxies) will be aware, the difficulty lies in finding a particular galaxy with some degree of certainty.

In Part Two, the catalogue section forms the bulk of this volume, and is split into three parts. The first part is a description of the central region of the Virgo Cluster and will be of interest to users of small aperture telescopes. Following an introductory section, visual observations of about 80 galaxies made by Webb Society members using instruments ranging from 12 × 80 binoculars to 18" reflectors are listed, together with drawings of a large sample of these galaxies. The second part of the catalogue is a collection of visual observations of 15 Abell clusters of galaxies, and contains observations of about 175 individual galaxies. The final part of the catalogue is a collection of visual observations of 13 groups or poor clusters of galaxies containing observations of about 120 individual galaxies.

1. The Spatial Distribution of Galaxies

The cataloguing of non-stellar astronomical objects began in earnest with the compilation of Messier which was completed in 1781[(1), (2)]. This catalogue of nebulous objects was prepared to assist astronomers engaged in sweeping for comets, which at the time, soon after Newton's theory of gravitation was formulated, were of great theoretical importance. Messier's original catalogue contained 103 entries many of which, especially some of those now known to be situated in the Virgo cluster, were originally observed by his compatriot Méchain. It is now known that 33 of the objects in Messiers original catalogue are galaxies, most of which are of visual magnitude less than 11. Almost all of the galaxies brighter than magnitude 10 in the northern sky are to be found in this catalogue which is used by many amateur astronomer as a standard list of bright deep sky objects containing nebulae and star clusters of every class. Modern authors have attributed several more objects to Messier bringing the total number of objects listed in the current catalogue up to 110.

The work of the Herschels in the late 18th and early 19th centuries represented a quantum jump in the mapping of space. Using various reflecting telescopes of up to 48" aperture, William Herschel compiled three lists of nebulae and clusters containing about 2,500 objects by systematic visual sweeps of the heavens. John Herschel extended the survey to the entire sky by observations carried out in South Africa and contributed a further 1,700 nebulae and clusters to the list. Although Herschel was keenly interested in the mapping of the astronomical neighbourhood of our planet in space, producing the first observationally based 3D map of stars in the galaxy, he was, of course, unaware of the true cosmological significance of the majority of the elements of his 'General Catalogue'. The concept of 'Island Universes' and the plurality of galaxies had to await the demonstration of the extragalaxtic nature of some of the nebulae in the early part of the 20th century by relatively accurate distance indicators. The greatness of their observational achievement is documented by the catalogue which, in revised form, is still in everyday use by the astronomical community. The final catalogue contained almost 5,100 nebulae and clusters and encouraged many observers to give their attention to this field and additional lists of objects soon appeared, many overlapping.

Due to the difficulty of visually observing faint nebulae in varying conditions, obscurities and misidentifications in all the catalogues became apparent. It was for these reasons that J.L.E. Dreyer, an astronomer who observed at Lord Rosse's great visual observatory at Birr Castle in Ireland, eventually published an updated and edited version of Hershel's General Catalogue - the New General Catalogue - in 1888[3]. The NGC contained positions and concise verbal descriptions based upon visual observation for 7,840 nebulae and star clusters. At about the time of the publication of the NGC, astronomical photography was becoming an important tool and it became apparent that supplementary catalogues would be needed to include the large number of new objects which were being discovered. Dreyer eventually published Index Catalogues to the NGC in 1895 and 1908[3] which, together with the NGC, contained a total of more than 13,000 entries. The method of compilation of the catalogue inevitably led to a hotch-potch of non-homogeneous coverage of the distribution of nebulae on the sky. Some areas were extremely well surveyed down to photographic magnitudes as low as 15 whilst other areas were poorly covered as revealed by modern wide field photographic surveys.

A further revision of the NGC was published in 1966 by Sulentic and Tifft[4] who cross-correlated the NGC with objects on the Palomar Sky Survey. This latest revision is invaluable to the modern visual observer, giving positions of epoch 1975, integrated photographic magnitudes, nebular/cluster type and a description of each object from its imprint on the POSS plates, as well as a large amount of other interesting information. Using the magnitudes quoted in the RGNC, the NGC appears as a non-homogeneous sample of galaxies down to about magnitude 15.5 where it is very incomplete. As was mentioned in the introduction, all 13,000 NGC and IC objects are plotted on the SAO atlas and graphically illustrate the clustering of galaxies at magnitudes brighter than about 15.5. Almost any chart of an area of high galactic latitude contains a fairly uniform sprinkling of nebulae - mostly galaxies. Most of the sheets also contain at least two or three knots of more than three or four nebulae. The NGC objects were first plotted in this way by R.A. Proctor in the late 19th century in the form of separate charts of the northern and southern celestial hemispheres, and the main features of the distribution of galaxies within the survey volume were already very clear. More complete surveys hardly affect the gross properties. Firstly, there is a vast desert of objects about 20° wide centred on the Milky Way. This is known as the Zone of Avoidance and the explanation of its existence is now obvious with the modern recognition of the shape of our galaxy, our position within it and of the extragalactic nature of most of the NGC objects. Quite simply, extragalactic objects of low galactic latitude are progressively dimmed by obscuration due to opaque clouds of dust which lie in the plane of the galactic disc. Indeed, more modern surveys of the distribution of galaxies have been used to yield data on the distribution of dust in the galactic plane. The second most pronounced feature of Proctor's chart is the great cloud of nebulae centred upon the Virgo Cluster which is now known as the Local Supercluster, and the general clustering of nebulae over the whole sky. Another gross feature is the relatively low density of nebulae in the southern galactic hemisphere, in a direction directly opposite to the local supercluster.

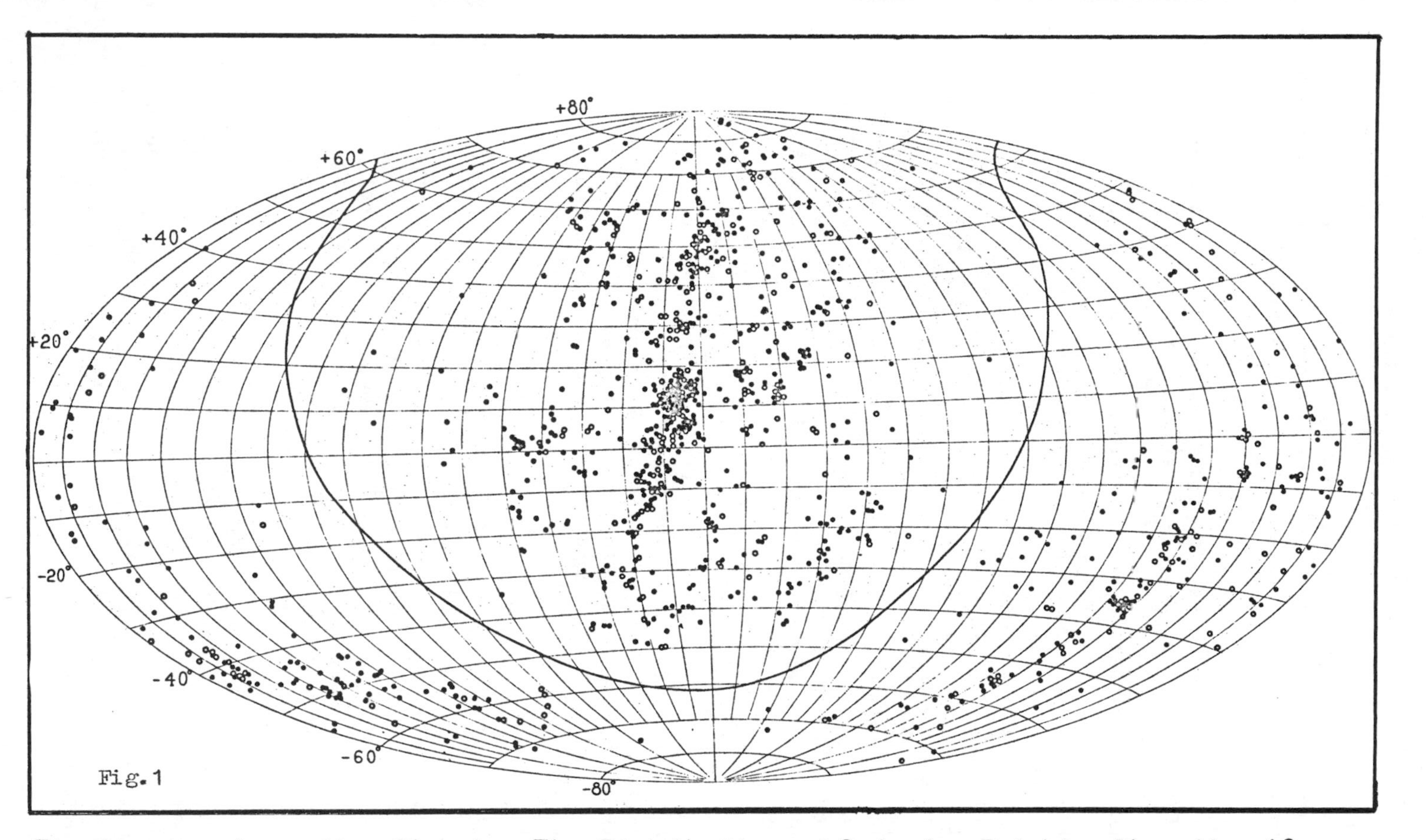

The Shapley-Ames Map Showing The Distribution of Galaxies Brighter Than Mag 13
(taken from Annals of Harvard College Observatory Vol.88,No2,1932)

Reproduced by kind permission of the Harvard College Observatory.

The lack of homogeneity of the NGC and IC survey of galaxies led to a series of surveys of their distribution on the sky which were intended to be as homogeneous and as complete as possible to various limiting magnitudes.

The first great survey dates from 1930 when the preparation of the Shapley-Ames Catalogue[5] of galaxies was begun at Harvard College Observatory. The latter catalogue was intended to contain positions, magnitudes and other classificational material for all galaxies on the entire celestial sphere brighter than the 13th photographic magnitude. Eventually 1249 galaxies were detected and measured on a homogeneous collection of plates, but it has been subsequently recognised that the survey is only complete to photographic magnitude 12.8. It is interesting to note that all but 61 of the 1249 galaxies originally appeared in the NGC, and that 48 of these are IC objects. Thus the NGC and IC survey is reasonably complete to this magnitude.

The distribution of the 13 mag. galaxies on the sky is shown in Fig. 1. The most prominent concentration is the local supercluster superimposed on an almost random sprinkling of more isolated galaxies. There are many knots or apparent clusters visible in both galactic hemispheres, the most notable in the southern galactic hemisphere being the Fornax cluster at about 03^h00, -40°, the largest of the nearby southern clusters of galaxies. Recall that for galaxies of average luminosity (Mabs = -20.0) the 13th magnitude corresponds to a distance of about 40 Mpc. Of course, some of the Shapley-Ames galaxies near the 13th magnitude limit are nearby dwarf systems, whilst others are distant giants. In 1964, the first revision of the Shapley-Ames catalogue was produced by de Vaucouleurs in the Reference Catalogue of Bright Galaxies[3] in which all available data on the Shapley-Ames galaxies is collected.

Since the original compilation of the Shapley-Ames catalogue, sky survey techniques have improved considerably with the use of large aperture widefield Schmidt telescopes, and the most recent surveys of deep space have been based upon the sky atlases thus produced. However, the earliest deep surveys used large photographic refractors. The first to go deeper than the 18 mag. survey was the 19th/mag. (400 Mpc) survey carried out by the Harvard investigators, using the 16 inch Metcalf refractor for the northern sky and the 24 inch Bruce refractor in South Africa for the southern sky, both having extremely wide fields enabling them to photograph up to 30 square degree areas at a time. The plate limit for the identification of stellar images was 18 mag., but galaxies could only be positively identified down to about mag. 17.6, the limit for completeness. The richness of the galaxy distribution is indicated by the fact that, typically, 4000 galaxies could be captured on one plate with a 3 hr. exposure. In all, about 600,000 galaxies were detected.

A more recent important survey made with a photographic refractor was carried out at Lick Observatory by C.D. Shane and his colleagues using the 20" Lick astrograph. The region of sky covered was limited by their horizon, but it is complete to about the 19th photographic magnitude, corresponding to a census of galaxies which reaches out to a distance of almost 630 Mpc and which includes about 1,000,000 galaxies. The compilation of the survey took about 12 years, and despite the modern Schmidt sky atlases, the resulting Shane-Wirtanen map of galaxy distribution remains one of the most important. In this deep survey, the Local Supercluster loses the contrast it held on the 13 mag. Shapley-Ames map due to the profusion of background objects. Some clustering is still discernible, for example the Coma Cluster stands out very prominently as a dense ball of galaxies near the north galactic pole, but the distribution now appears almost random except for a very interesting, possibly subjective, filamentary structure. We shall refer to it again in a later section when we discuss the nature of galaxy clustering.

Modern survey astronomy is based to a very large extent on the great Palomar Observatory - National Geographic Society Sky Survey (POSS) - carried out in the 1950's using the newly erected 48 inch Schmidt telescope, still the (co-) largest of its class. The original survey was of the entire sky visible from Mt. Palomar - down to a declination of about -27°, but it is being supplemented in the southern hemisphere by the UK 48 inch Schmidt at Siding Spring Australia and the ESO 40 inch at La Silla in Chile. The original survey covered the sky area with 879 pairs of photographic plates. The plates are of areas about 6° square with a non-vignetted field of about 5.4^{o} in diameter, and the fields were selected to allow for at least 0.6^{o} of overlap along adjacent edges. The fields were photographed twice on photometric nights using red and blue sensitive plates with limiting photographic stellar magnitudes of 21.1 for the blue series and 20.0 for the red series. The UKSTU-ESO southern hemisphere survey extends to a somewhat fainter limiting magnitude than the original POSS due to the increased sensitivity of modern emulsions. Because of the redshift of light of very distant galaxies, the red series of plates is of more use for deep galaxy surveys in that the peak intensities of normal distant galaxies are shifted well into the red end of the spectrum for red-shifts of the order of 0.2 and above.

Two great galaxy catalogues have been prepared using the POSS. The first is the great survey carried out by Zwicky and his collaborators at Mt. Palomar in their Catalogue of Galaxies and Clusters of Galaxies[6] - the CGCG. The catalogue contains positions and magnitudes for about 30,000 galaxies and about 10,000 groups and clusters of galaxies for galaxies down to about magnitude 15.7, but it is probably only reasonably complete down to about the 15th magnitude. A similar but more detailed catalogue was prepared under the supervision of B.A. Vorontsov-Velyaminov at the Sternberg Astronomical Institute of the University of Moscow for galaxies down to about magnitude 15.1. The catalogue is called the

Morphological General Catalogue[7] - (The MGC), and contains detailed notes on the appearance of over 35,000 objects on the POSS.

The number of galaxies which appear on the POSS plates is almost uncountable without using automated techniques, and therefore the coverage of the Shane-Wirtanen 19th magnitude survey has not yet been bettered. Detailed surveys have, however, been completed for small areas of the sky to larger limiting magnitudes. One of the most important of these is the 20.5 magnitude survey of an area of sky on a single $6^{o} \times 6^{o}$ Schmidt plate by K. Rudnicki on which about 10,000 galaxies were counted. At such extreme distances (of the order of about 1000 Mpc) the clustering of galaxies is not as conspicuous as it appears on the 13th and 15th Mag. maps due probably to the superposition and overlapping of many loose clusters at intermediate distances, just as in the Shane-Wirtanen map.

The recent use of computerised digital techniques for plate scanning and galaxy identification promises to revolutionise work on galaxy distribution. One of the latest automatic plate measuring machines is appropriately known as COSMOS and computer analysis of its digital output is able to distinguish faint stars from almost stellar galaxies with an accuracy approaching 95%. The gain in information extraction can be roughly gauged from the fact that the Shane-Wirtanen survey to the limiting magnitude of the Lick astrograph counted an average of about 50 galaxies per square degree, whilst the Rudnicki map contained an average of 348 galaxies. The COSMOS machine is able to count about 2000 galaxies per square degree down to the limiting magnitude of an average 48 inch Schmidt IIIaJ plate. No results are yet available for a complete survey of the sky down to about the 20th magnitude, but the speed and accuracy afforded by the new techniques will probably one day make such a survey feasible.

The great detail contained in the Shane-Wirtanen and Rudnicki maps requires a detailed statistical analysis in order to extract significant detail on the large scale distribution of galaxies in the universe. The results of the analysis will be discussed in a later section after we have described the main catalogue upon which this volume is based - the Abell Catalogue of Clusters of Galaxies.

2. The Abell Catalogue of Rich Clusters of Galaxies

In the previous section, the history of the surveys of the distribution of galaxies in space was outlined with the emphasis laid upon limiting magnitude as an indication of the spatial depth of the surveys. A weakness of this approach is the fact that galaxy luminosities vary by a considerable amount, the variation of the numbers of galaxies with a given absolute magnitude - the luminosity function for galaxies - being rather uncertain. Suppose that one were to use clusters of galaxies as cosmological probes. Then hopefully, one would be able to define certain characteristic properties of clusters which one could use as distance indicators. In this way, the three dimensional distribution of clusters of galaxies could be used as a probe of the matter distribution in the universe. For example, suppose that the variation of the number of member galaxies of a cluster with absolute luminosity was the same for all clusters, that is, that the luminosity functions of clusters had a characteristic form. Then one could use, say, the apparent magnitude of the n-th brightest member of a cluster as a distance indicator. This type of reasoning is open to objection, but was used as a working hypothesis by G.O. Abell in 1957 in his classic paper[8] which contained a catalogue of 2712 rich clusters of galaxies in the POSS sky area.

Some important cosmological work of Sandage prior to and subsequent to the appearance of Abell's paper seems to justify the assumption of a universal luminosity function for clusters used by Abell. For example, the apparently small variation in absolute magnitude of first ranked cluster galaxies is well known. Similar small variations were also established for the 3rd, 5th and 10th ranked galaxies in clusters by Sandage and his co-workers. The variation of other cluster parameters with redshift, also seem to require a 'universal' rich cluster model for their explanation.

Relatively bright, nearby clusters of galaxies are easy to distinguish against the background of fainter galaxies except that confusion can arise between faint cluster members and background objects. The probability that a compact distribution with a relatively large number of galaxies of similar appearance and brightness is a real association rather than a statistical fluctuation is rather high, thus rich compact clusters are easy to recognise. However, for small clusters, say containing about four or five members, the statistical results are not so compelling, and the demonstration of real association in space becomes extremely difficult in the absence of distance information such as redshifts. A very good example of this is the classic case of Stephan's Quintet, a compact group of galaxies in Pegasus. All five galaxies are extremely close together on the sky but one of the five (NGC 7320) has a much smaller redshift than the others. Various explanations of the discordant redshift are possible[9]. It is therefore important to decide if all five galaxies do lie at a similar distance since it seems most likely that NGC 7320 is a foreground object. Years of academic argument based upon the most sophisticated astronomical techniques has failed to rule on the relative distances within the group.

Before Zwicky's work for the CGCG, only a couple of dozen clusters of galaxies were catalogued, most being due to Shapley. These were chosen in an almost random manner. Zwicky's survey eventually increased the number of catalogued clusters of galaxies to about 10,000 but it was a very inhomogeneous survey, ranging from groups of four or five galaxies (therefore probably contaminated by accidentally projected groups) to large, sprawling complexes spread over many degrees of sky, the latter being open to heavy contamination by background objects and thus having low probabilities of being physical associations without the detailing of individual redshifts. For these reasons, Zwicky's catalogue of clusters is little used today. By adopting a formal definition of a rich cluster of galaxies, Abell was able to compile a homogeneous catalogue of rich clusters down to near the limiting magnitude of the red series of the POSS.

Clusters of galaxies became very difficult to delineate on deep plates due to the very heavy distribution of galaxy images, and as we noted, only rich relatively compact clusters have good contrast against a crowded background sky, and high statistical probability of being real physical associations. For this reason, and to ensure a uniform sample, Abell was led to the following definition of a rich cluster of galaxies. (Note that this definition is a function of the POSS, for example if his criteria was used to define a rich cluster on the NGC + IC sample plotted in the SAO atlas, hardly any groups except perhaps the Coma cluster would qualify as rich clusters.)

1) Richness. A rich cluster must contain at least 50 members not more than 2 mag. fainter than the 3rd brightest member. Here, the 3rd brightest member was chosen as a reference point to try to avoid errors due to possible contamination by bright foreground galaxies.

2) Compactness. A rich cluster should be sufficiently concentrated so that the 50 members fainter than the 3rd brightest member lies within a distance R of its centre. Of course the radius R corresponds to an angular distance on the sky which will decrease with increasing distance or redshift. One could estimate the cluster redshift (z) using the apparent magnitude of the 10th brightest member. The disc size used by Abell in his examination of the POSS was $(9.2 \times 10^5/cz)$ mm diameter where c is the speed of light - $(3.0 \times 10^5$ km sec$^{-1})$, cz being the redshift recession velocity. The corresponding characteristic cluster diameter assumed was therefore 6.0 Mpc, defined by the angular scale: 67.1 seconds of arc per mm of the POSS, using $H_o = 50$ km sec^{-1} Mpc^{-1}. The results are relatively insensitive to this working rule since, in practice, most galaxies are contained in a core of much smaller dimensions.

3) Distance Criteria. A cluster must be sufficiently distant so that its apparent radius is contained in a single POSS plate, perhaps overlapping only with one other plate. This excludes many close clusters such as Virgo which would require several plates for complete coverage. An upper limit was set by the richness criterion - one must be able to distinguish galaxies 2 mag. fainter than the 3rd brightest galaxy in a rich cluster. Since the red plate limit is 20.0 mag. and galaxies fainter than 19.5 mag. could not be differentiated from stars with complete certainty, this implied a limit through the requirement that the 3rd brightest member be at most of mag. 17.5. Abell adopted a lower distance limit corresponding to a redshift velocity of 6000 km sec^{-1}

(or 120 Mpc for $H_o = 50$) and an upper distance limit corresponding to a redshift velocity of 60,000 km sec^{-1} (or 1,200 Mpc).

4) Galactic Latitude. At very low galactic latitudes, interstellar obscuration blots out all that lies beyond it. Even at moderately low latitudes where the obscuration may be relatively low, and galaxies shine through the thin veil, star densities can be so high as to render cluster analysis impossible through the teeming multitude of Milky Way stars. Because of this effect, Abell was forced to limit his survey to regions of varying minimum galactic latitude with longtitude. On average the survey extends north of galactic latitude 35° and south of galactic latitude -35° for the POSS region.

In the preparation of the catalogue, Abell included quite a few clusters of galaxies which did not fulfill his definition in all its requirements. For example, quite a few poor clusters with less than the requisite 50 members fainter than the 3rd ranked member were included. These fall into his richness class 0 (see below). Also, quite a few clusters beyond the Milky Way are also included in the complete catalogue of 2712 clusters. In all, 1682 clusters included in the catalogue fitted the definition of rich cluster, and these formed the statistical sample he used for his analysis of the spatial distribution of rich clusters of galaxies.

All the clusters in the catalogue were categorised into varying degrees of richness and distance. Richness is a measure of the total population which is taken to be an increasing function of the number of galaxies less than two magnitudes fainter than the 3rd brightest member. The populations in this magnitude interval ranged between 30 (the minimum necessary to allow entry into richness class 0) and over 300 for a richness group 5. The richest of all the clusters in Abell's catalogue is the cluster Abell 665. It should be noted, however, that the correspondence between Abell's definition of richness and the natural measure - total population, is dependent on the 'universality of the luminosity function for clusters'. For example if the function differed from cluster to cluster, a change in the slope of the bright end of the integrated luminosity function might drastically affect the richness class but might not affect the total population.

A cluster with a deficiency of brighter galaxies will have a low number of galaxies in the magnitude interval m_3 to $m_3 + 2$, implying a low Abell richness class, but may have a large population of relatively faint galaxies. For example, the cluster Abell 2029 is such a case, having Abell richness class 2 but a total detectable population of 1600. A cluster with an excess of brighter galaxies will have a high number of galaxies with magnitudes in the interval by m_3 to $m_3 + 2$ but may have relatively few fainter members, and hence a relatively low total population. The example here is the cluster Abell 665, which has richness class 5, but a total detectable population of about 1000 members.

Recall that the distances of rich clusters were estimated through m_{10}, the apparent magnitude of the 10th brightest cluster member. The distance classes used to compare cluster distances are based upon m_{10} as in the table below:-

Table 1

Abell Distance Group	Magnitude of 10th brightest galaxy
1	13.3-14.0
2	14.0-14.8
3	14.9-15.6
4	15.7-16.4
5	16.5-17.2
6	17.3-18.0
7	18+

Abell Richness Group	Number of galaxies in the interval m_3 to $m_3 + 2$
0	30-49
1	50-79
2	80-129
3	130-199
4	200-299
5	300+

It should be noted that the distance estimate based upon M_{10} also depends on the cluster luminosity function, and although distance is clearly an increasing function of Abell distance class, a cluster with an excess of bright galaxies will appear closer than one with only a few; in short, both the Abell distance and richness classifications are open to criticism, but serve well as working indications.

3. The Distribution of Clusters of Galaxies (see Figure 2A, page 36)

Because Abell's catalogue was the first homogeneous complete survey of a well-defined sample of clusters of galaxies, it was the first on which sensible statistical analyses could be performed on the distribution of clusters in the universe. Zwicky had discussed the distribution of clusters in certain areas of the sky, but his main conclusion, that their distribution is entirely random, is contradicted by most modern authors. Shane and Wirtanen had noted possible clusters of clusters on their Lick survey plates. Indeed, some second order clustering is apparent even on plots of the NGC and IC samples of galaxies. For example on the SAO, there is a chain of groups and clusters which stretches from the cluster Abell 262, through the NGC.507 group to the Pisces Group (The NGC.383 group) across some 15° of the sky. Another chain of groups and clusters stretches from the Perseus Cluster (Abell 426) through the NGC.1130 group to the cluster Abell 347, again about 15° in apparent size. Zwicky had recognised such groups but regarded them as mere statistical fluctuations. Modern results suggest such chains are physical associations on gigantic scales.

Using his statistical sample Abell was able to derive the following conclusions.

1) Clusters are distributed approximately uniformly in depth. This conclusion was reached by having a subdivision of clusters into distance classes. A plot of the logarithm of the cumulative total populations of the distance classes against distance classes yielded a good approximation to the theoretical straight line with slope 0.6 expected for uniformity.
2) There is a highly significant non-random surface distribution of rich clusters of galaxies for the various distance classes which suggests a second order clustering of clusters of galaxies with a characteristic scale of about 90 Mpc which is suggestive of a typical size for a second order cluster (or supercluster).

Since Abell's pioneering work, many statistical analyses of the distribution of clusters have been performed, mostly based upon his catalogue. Most authors verify Abell's conclusions. For example Rood(10) detected five superclusters in distance groups 0-2, and about 40 probable superclusters in distance groups 3-4, most superclusters identified being binary or triplet clusters. He also demonstrated that the superclusters are very probably randomly distributed, a factor which demonstrates that 3rd order clustering probably does not exist, implying that the universe is homogeneous when considered on large enough scales.

The conclusions on 2nd and 3rd order clustering derived from analysis of Abell's catalogue, with additional good redshift information mentioned above, is confirmed by classic analyses of the distribution of galaxies undertaken by P.J.E. Peebles and his co-workers at Princeton (11). Their work is mainly based on the 15 mag. Zwicky survey, the Shane-Wirtanen study and the Rudnicki galaxy counts using a method called correlation analysis. Correlation analysis is a statistical method of analysing the tendency of galaxies to cluster. For example, one can analyse the tendency of galaxies to cluster in n-tuplets through the n-point correlation function defined roughly as follows. One counts the number of n-tuplets of galaxies which have an angular diameter in some given range in a survey area. This number is then divided by the number of n-tuplets expected for a distribution of galaxies which is uniform in depth, but otherwise random across the sample area. The ratio of the observed number to the expected number is a measure of the randomness of the sample. If the observed distribution is random, the ratio would be exactly 1, so the n-point correlation function is defined as the ratio minus 1 and an n-point correlation function of zero describes a perfect randomness for the distribution of n-tuplets of the given diameter. Peebles et al. have demonstrated that after corrections due to their differing depths, the 2-point correlation functions on the 15th mag. Zwicky Survey and the 19th Mag. Shane-Wirtanen survey are almost exactly the same, whilst the correlations for the 20.5 mag. Rudnicki sample is approximately the same. Note that as depth of survey increases, the task of computing the correlation becomes increasingly tedious as the surface density of galaxies increases. This result indicates that statistically, galaxy clustering in very deep space is identical with clustering on a more local scale.

Plots of the 2 point correlation functions for the three surveys mentioned above against angular diameter, provide evidence for super-clustering. For example, if plotted on logarithmic scales, the logarithm of correlation decreases linearly with the logarithm of the angular separation quite smoothly until, at angular separations which decrease linearly with increasing distance, the slope changes rather abruptly. Thus the tendency for galaxies to cluster decreases very sharply at a certain characteristic scale. Using the fact that the break occurs at about 10° for the 15 mag. survey yields a characteristic scale of about 20 Mpc if we assume that the 15 mag. survey corresponds to a depth of 100 Mpc. The analysis of galaxy distributions indicates that galaxies cluster on a scale up to about 20 Mpc, but that clustering on larger scales is probably random. Further calculations based upon the 3 and 4-point correlation functions, which become extremely tedious to carry out on the Shane-Wirtanen map, broadly verify the above conclusions.

Two recent studies have shed much light on the second order clustering of nearby clusters of galaxies. The first of these is due to S.A. Gregory and L.A. Thompson(12) and it demonstrated observationally that superclustering does indeed occur on scales up to about 60 Mpc. The authors obtained redshifts for galaxies in fairly rich groups of galaxies of around photographic magnitude 15, which lie between the Coma cluster Abell 1656 and the rich cluster Abell 1367 in Leo. The chain of groups and clusters can be traced through its NGC/IC galaxies across sheets 57 and 39 of the SAO atlas and rich intermediate subgroups are labelled from A1367 and the NGC.3937 group, the NGC.4065 group and the NGC.4213 group, all well within range of the well-equipped amateur. The system stretches across about 20° of sky at a redshift of about 7000 km sec^{-1}, corresponding to a size of about 40 Mpc.

The second approach has provided compelling evidence for a lot more sprawling superclusters in relatively nearby space. G. Chincarini and H.J. Rood(13) used a homogeneous, all-sky survey of a large sample of Sc galaxies, prepared by V.C. Rubin and her collaborators for an unrelated purpose. It was assumed that the spatial distribution of Sc galaxies is in good correspondence with the distribution of all galaxies. The sample of Sc galaxies lay in the magnitude range 14.0 to

15.0, and correspond to a redshift range of 4200 to 6700 km sec^{-1}, or to distances of 84 Mpc to 130 Mpc if the Sc galaxies are ideal 'standard candles' with absolute luminosity -21.24. Of course, galaxies are observed outside this distance scale because of the spread in absolute luminosity with a measured dispersion of about 0.3 mag. Comparing the observed luminosity function of the sample with a possible Gaussian distribution, peaks are evident in the function at redshifts of about 53000, 7300 at 101900 km sec^{-1}, that is, excess numbers of galaxies appear at these redshifts. Since the sample was chosen to give a homogeneous coverage in terms of galaxies per square degree across the sky (outside the zone of avoidance), the result is suggestive of galaxy clustering at these redshifts. The latter redshifts are very near those of the superclusters associated with the Abell clusters A426 (Perseus), A1656 (Coma), A2151 (Hercules).

Another peak in the galaxy count appears at a redshift of 2900 km sec^{-1} corresponding well with the redshift of the group of galaxy clusters dominated by A1060 (Hydra I) and the Centaurus cluster. The superclusters can be traced across the sky by plotting the positions of the sample galaxies in a small range of redshifts. In this way, the enormous angular sizes of the superclusters are revealed. For example, Gregory and Thompson's map of the A1656/A1367 supercluster is confirmed in essence, but appears even larger, stretching south in galactic latitude to the cluster A779. The clusters A1185, A1213 and A1228 all appear to be associated with this cloud, which can be traced out to an angular dimension corresponding to a physical size of about 200 Mpc, with a radial depth of at least 50 Mpc.

Another supercluster of galaxies is named after the Perseus cluster A426, the association extending irregularly across most of the sky and having a diameter of about 200 Mpc. The Abell clusters A426, A347 and A262 indicate its densest concentrations. The Sc distribution also links the group of clusters around the Hercules cluster A2151 (A2151, A2152, A2147) at an average distance of about 220 Mpc, with a group of clusters around the cluster A2199 (A2199, A2197 and A2162). Moreover, the structure may even extend over most of the sky, to include the clusters A2634 and A2666 which lie at a similar distance.

The Chincarini and Rood investigation described above suggests that the nearby rich clusters of galaxies form concentrations in vast clouds or superclusters separated by immense voids. The distribution of the superclusters turns out not to be in concentric shells of galaxies, as the above descriptions may suggest, but rather in a three dimensional system of cells. The galaxies populate the areas near the boundaries, with strings distributed in depth being revealed by polar plots of the Sc sample with redshift in the radial direction.

The detail of this nearby large scale distribution of clusters of galaxies is almost all observable to some extent by the well equipped amateur. Many amateur astronomers possess telescopes of 12 inches aperture or more and can obtain a lifetime's enjoyment in the observation of galaxies down to about photographic magnitude 15.0, or even fainter for those lucky enought to have access to equipment of aperture from 16 inches to 24 inches. The number of galaxies to this magnitude (at least 30,000) provides a virtually inexhaustable source of personal discovery and wonder. Most of the 30,000 have never been observed visually, and thus one can experience, in this overcrowded world, the satisfaction of being the first to contemplate since their creation, the almost infinitely dimmed light of a myriad suns. Having discussed the distribution of clusters of galaxies in space, the next chapter will discuss the physics of individual clusters from an evolutionary and morphological point of view.

3. THE PHYSICS OF RICH CLUSTERS OF GALAXIES

1. The Formation of Clusters and Superclusters

Current cosmological theory is dominated by the Hot Big Bang models of the universe which rest on three very strong observational foundations. These are the Hubble redshift-distance relation for distant galaxies, the 2.7°K universal background microwave radiation and the observed cosmic helium abundance. The interpretation of the redshift-distance relation comes from noting that distant galaxies are observed deep in the past. It then follows that since light emitted in the past appears redshifted to us, the universe must have been smaller in the past by a factor (1 + z) at an epoch corresponding to redshift z. Straight-forward extrapolation of this conclusion suggests that the whole universe must have been compressed into an extremely small size at times of the order of H_o^{-1}.

In the mid 1960's, Penzias and Wilson established the existence of a background microwave radiation of wavelength about 1 mm emanating isotropically from the whole sky, with a blackbody energy distribution corresponding to a radiation temperature of 2.7°K. Such a universal background radiation is a natural consequence of big bang cosmology. Suppose that the universe were filled with radiation with temperature T. Since T is proportional to the peak radiation frequency, it is inversely proportional to wavelength and hence proportional to (1 + z). Therefore, at epochs corresponding to large redshifts, the radiation temperature must also have been large, and at a suitable redshift, about z = 1000 when T = 2700°K, the photon energy would have been sufficient to ionize completely most of the matter in the universe. Neutral matter is virtually uncoupled to radiation, but ionized matter is strongly coupled. Thus at z = 1000, matter and radiation would have been in thermal equilibrium at a temperature of about 3000°K. Thus the 2.7°K microwave radiation is the redshifted, cooled ghost of a hot early universe of radiation and plasma.

The densities of both matter and radiation increase rapidly with increasing redshift. The matter density increases in the obvious way as $(1 + z)^3$, but according to the Stephan-Boltzmann law, the radiation increases in density as $(1 + z)^4$, that is, the radiation density changes (1 + z)-times as quickly as the matter density. At the present epoch, the mean density of matter is estimated to be of the order of 10^{-30}gm/cm^3, whilst the density of radiation, estimated from the Stephan-Boltzmann Law, is about 10^{-33} gm/cm^3 for T = 2.7°K. Therefore the density of matter is about 10^3 times that of radiation at the present epoch, and it follows that at a redshift of about 10^3, the radiation density of the universe exceeded that of matter. It appears that, in our universe, matter and radiation decoupled at about the same epoch that matter became the dominating factor determining cosmic evolution. The early, radiation dominated, slowly cooling era of the universe, lasted for a relatively short time (about 10^9 years) compared with the Hubble time. In that short time, the seeds of the structure of the universe of matter were already sown, and remained approximately constant during that chaotic period. As the mixture of high energy elementary particles and radiation cooled, progressively more complex atomic nuclei 'froze out'. The conditions at early times were ideal for the formation of helium in about its observed cosmic abundance. In fact, helium is so plentiful in the universe, that the big bang is probably the only means of creating it in sufficient quantities: its formation inside stars, and subsequent distribution into the void by supernova explosions being much too slow a mechanism.

It is within the above cosmological framework that the formation of galaxies and clusters of galaxies will be discussed. The theory which will be outlined can account for some of the gross properties of the observed distribution of matter and some of the observed morphological characteristics of individual clusters whose description is the main aim of this chapter.

Two conflicting forces are everywhere at play in the universe. These are gravitation, which tends to concentrate local density enhancements, and hydrodynamic pressure, which tends to resist such compression. On the universal scale, the outcome of the conflict will decide if the universal expansion will slow down (gravity wins) or go on for ever (pressure wins). The pressure is a measure of the kinetic energy imparted to the cosmic fluid during the big bang, and the ratio of the gravitational potential energy to the kinetic energy of expansion is denoted by the symbol Ω. For $\Omega > 1$, the local density of matter is sufficient eventually to overcome expansion, whilst if $\Omega < 1$, the kinetic energy of expansion is sufficient to ensure continual expansion.

Various methods exist for making informed guesses about the value of Ω: most suggest $\Omega < 1$. The most sensitive indicator, the cosmic deuterium abundance, suggests that Ω is about 0.1. On the local scale, the conflict serves to regulate the formation of protogalaxies and protoclusters. A certain mimimal total mass is necessary to exert sufficient gravitational self-attraction for collapse to take place. This is called the Jeans Mass. For masses less than the Jeans Mass, local density enhancements oscillate rather like acoustic waves.
The Jeans Mass varies with cosmic epoch in big-bang cosmologies. In the radiation dominated era, the critical mass increases from very small values to a maximum of about 10^{18} solar masses at the epoch of recombination. The latter figure is much larger than the mass of large superclusters of galaxies. Density enhancements on smaller scales than this must have become unstable, to collapse at earlier epochs, but would

eventually have become stable in an oscillatory sense. An important damping mechanism is effective in the radiation era, and acts to smear out small scale density enhancements. The mechanism is known as 'photon viscosity' and it can be shown that it will effectively erase density enhancements with masses less than those of clusters of galaxies.

At the epoch of recombination, the Jeans Mass decreases very rapidly to a value of about 10^5 solar masses, the mass of typical globular star clusters, and then decreases steadily with cosmic time. Note that on this model, the only density perturbations which can have survived through the radiation era will be those with masses of the order of large clusters of galaxies, smaller ones having been smeared out by photon viscosity. The important factor now becomes the rate of collapse of the protoclusters after the epoch of recombination. One can show that the relative strength of a weak density enhancement over the background density ($\delta\rho/\rho < 1$) increases like $(1 + z)^{-1}$, with cosmic epoch for redshifts z with $z > \Omega^{-1} \simeq 10$. For smaller redshifts, the weak enhancements grow very slowly. Those density enhancements with $\delta\rho/\rho > 1$ will grow rapidly, allowing internal density enhancements to collapse into individual stars and galaxies. It therefore follows that the last proclusters to collapse will be those with $\delta\rho/\rho \sim 1$ at a redshift of 10, enhancements with $\delta\rho/\rho$ greater than 1 will collapse earlier. Therefore the smallest enhancement capable of collapse will have to reach $z = 10$ with $\delta\rho/\rho$ about equal to 1, and these will form the bulk of the objects we observe today, because one would expect a primeval spectrum of fluctuations with larger numbers of small fluctuations on statistical grounds. It has been estimated that the objects reaching $\delta\rho/\rho = 1$ at $z = 10$ will complete their initial collapse by $z = 5$, labelling the epoch of galaxy formation in the universe.

The scale of protocluster masses is sufficiently large to encompass the possibility that superclusters may be very large clusters with internal subclustering. Computer simulations of cluster collapse indicate that in the early phases of collapse, very strong subclustering is to be expected even from an initially uniform protocluster. Subsequent evolution involves the continual blending of smaller subclusters until a symmetric distribution is attained for a relaxed cluster. The observed uniform distribution of superclusters is in accordance with the expected uniform spatial distribution of protoclusters whose distribution should broadly follow from the assumed isotropy and homogeneity of the background cosmology.

2. The Physics of Individual Clusters

As a protocluster collapses, chaotic, turbulent motions are to be expected within its bulk which will inevitably produce strong, rotating density enhancements on all scales, with a range of masses exceeding the local Jeans masses - the latter being dependent on local temperatures and pressures. The small scale density enhancements will collapse rapidly, generally into small groups of galactic masses. Within the protogalaxies, stars will condense and collapse to a stage where fusion reactions can begin in their centres. Much gas and dust will remain uncondensed in-between the infant galaxies and stars. In those galaxies with a larger proportion of matter used in star formation, elliptical galaxies will form, and early generations of hot, massive, rapidly burning stars will utilise the bulk of free material and sweep out any remaining material by mechanisms such as supernova explosions and local radiation pressure. The majority of stars will be of low mass, and these still remain - the slowly evolving red population of elliptical galaxies and bulge population of disc galaxies. Those galaxies which do not use up a large proportion of their nascent matter in star formation will evolve into disc galaxies, where rotation acts on the remaining gas and dust, flattening it into the disc component. Gravitational instability within the disc may produce spiral shaped density enhancements, which rotate relatively slowly around the disc. The differentially rotating matter within the disc periodically passes through the density enhancements, and is compressed, triggering the bulk of hot blue stars which trace the density wave matter we observe in spiral galaxies. Some disc galaxies may be deficient in gas and lack distinctive blue spiral arms. These are the SO galaxies.

The initial motions of galaxies within a cluster are determined by the random turbulent flows within the collapsing protocluster, and by gravitational encounters between protogalaxies. An asymmetric protocluster will collapse preferentially along its minor axes, possibly producing oblate or prolate ellipsoidal clusters. Collapse will also proceed more quickly within the central regions of the protocluster than in more outlying regions, possibly producing a relatively dense core surrounded by a slowly collapsing halo of galaxies.

It is very probable that most of the clusters and superclusters observed today are still undergoing some form of dynamical evolution, for calculations indicate that there appears to be insufficient luminous mass in almost all clusters to bind themselves gravitationally in dynamic equilibrium. For example, in the Coma cluster (Abell 1656) it has been calculated that the matter observed in visible galaxies is only about 10% of that necessary to balance the observed velocity dispersion ('galaxy pressure'). Many explanations of such mass discrepancies have been proposed, some of which will be mentioned below, but it is possible that the main result - lack of dynamic equilibrium - should be taken at face value for many clusters. For example, the Hercules cluster (Abell 2151) is very irregular, and exhibits much subclustering. It may even be that A2151 and its companion A2152 (the main constituents of the Hercules supercluster), represent subcondensations in the still collapsing supercluster. Another example is Abell 426, (the Perseus cluster) which has the largest 'missing mass' problem of all known clusters, only containing 5% of the mass necessary for gravitational equilibrium. This cluster is highly asymmetric, most of its galaxies being distributed in a prolate chain of galaxies about 750 Mpc in length. The possible dispersive gravitational instability of clusters of galaxies is hard to accept, because if many of the constituent galaxies really do have velocities which exceed the escape velocity from the cluster, one would expect the clusters to have dispersed by now, since typical cluster crossing times are much less than the age of clusters. One is forced to fall back upon the 'missing mass' solution and/or to use the non-equilibrium arguments used by many authors which invalidate the use of the virial theorem upon which the above mentioned calculations are based.

After the galaxies have crystallised out of a protocluster, one can expect a large amount of non-luminous gas to be left uncondensed. According to the classic work by Gunn and Gott(1), most of this gas can be accreted onto cluster galaxies until the material is almost all exhausted. One can assume that the remaining gas will attain a very high temperature due to its gravitational compression in collapsing into the cluster core. Typical gas temperatures can be estimated by assuming that gas particles have similar velocities to galaxies, and this yields a temperature proportional to the velocity dispersion of the galaxies. Typical temperatures are of the order of 10^8 °K. Once the gas becomes extremely hot, accretion onto galaxies will cease since atoms will easily exceed the galactic escape velocities. Because of its extremely high kinetic temperature, the intracluster gas will radiate at X-ray wavelengths by the thermal acceleration mechanism. Such radiation has been observed in many clusters using satellite borne X-ray telescopes, and X-ray maps reveal the distribution of the hot radiating gas through diffuse X-ray sources concentrated around cluster centres. In the case of Abell 1656, the mass of hot gas required to produce the observed X-ray luminosity is insufficient to solve the missing mass problem. There is reasonable agreement between the theoretical proportionality of X-ray temperature and velocity dispersion in X-ray clusters. For example Mitchell et al.(2) have produced a convincing linear plot using the clusters A426, A1656, Centaurus, A401, A106, A1367 and A2256.

The gaseous intracluster medium can also be observed at radio wavelengths. The hot gas itself is a source of radio emissions and can be detected in several clusters as a large diffuse 'cluster source'. For example, in Coma, A1656, the cluster source is known as Coma C. A similar diffuse source was associated with the Perseus cluster A426, (3C 84B) by Ryle and Windram, although recent Westerbork observations by Gisler and Miley(3) have failed to detect it. Many isolated radio galaxies have twin lobes which straddle an optical galaxy symmetrically. The usual interpretation of this morphology is that twin plasmoids have been blasted out of the galactic core along the spin axis of the parent galaxy. If a similar phenomenon were to happen in a cluster galaxy in orbit through the intracluster medium, one would expect an interaction between the radio jets and the relatively static medium, which would produce a 'sweptback' morphology. This effect has been observed in many clusters such as Abell 1314, A779 and A347 amongst the nearby Abell clusters, and the radio sources are known as 'tadpole' or head-tail' galaxies[4]. The giant SO galaxy NGC 4889 in A1656[5] is a head-tail source, and the galaxy NGC 1265 in A426[6] is one of the most spectacular of all. In examining radio maps of the latter, one can naively visualise a 3-dimensional "contrail" trace of the orbit of the galaxy through the cluster. Estimates of the mass of intracluster gas in A426 from the above interaction are in rough agreement with the X-ray based estimates in concluding that the medium is insufficiently massive to explain the missing mass problem.

The intracluster medium is important in its interactions with galaxies via the 'ram pressure stripping' theory of Gunn and Gott. During its orbit around the centre of mass, a cluster galaxy will pass through the intracluster medium, and any gas present within the galaxy will experience a braking pressure due to its interaction with the intracluster medium. This pressure was termed 'ram pressure' by Gunn and Gott who showed that any passage through the relatively dense concentrations expected to be present near the cluster centre would probably be sufficient to strip most of the gas from rapidly moving galaxies. In this way, spiral galaxies would be progressively denuded of gas, and be converted into SO galaxies. Observational evidence to support this is mounting. For example rich, highly centrally condensed clusters, like A1656 and A426, are almost entirely populated by E and SO galaxies. The spirals which remain are those termed 'anaemic' by Van den Berg, possessing rather uniform spiral arms lacking the patchy emission regions and bright blue stars which characterise more isolated spirals. An outstanding example is NGC 4921 in A1656. Some of the bright spirals in the Virgo cluster also have this characteristic, a good example here is M90. Possible observations of ram pressure stripping in action include a deep photograph of faint blue filamentary structure extending from the disc of the anaemic spiral, NGC 3312[7] in the Hydra I cluster (Abell 1060), an X-ray cluster in the southern sky. A consequence of ram pressure stripping of interstellar gas from cluster galaxies is that the intracluster medium should be enriched with heavy atomic species. The detection of an X-ray emission line of highly ionised iron in the X-ray spectrum of the Virgo cluster by Ariel V instruments[8] seems to be good evidence for such enhancement.

Work on the relative numbers of SO galaxies in X-ray clusters has indicated that SO galaxies are relatively more abundant in clusters that have a large velocity dispersion, and would therefore be expected to have a highly efficient stripping process. It was also found that the abundance decreases to a normal level in outlying regions where stripping would be expected to be less efficient. In an attempt to understand the evolutionary histories of these galaxy types S.E. and K.M. Strom[9] have recently performed an extensive series of observations of E and SO galaxies in the clusters A1656, A1367, A426, A1228, A2151 and A2199. These encompass a range of cluster types, from rich concentrated sprial-poor clusters (A1656 and A426), to spiral dominated irregular clusters such as A2151. These workers conclude, for example, that the distribution of SO galaxies in A1656 lends support to the ram pressure model. An examination of the colours of edge-on SO systems indicated that relatively blue SO's are situated preferentially in the outer regions of the cluster. They also showed that the outlying SO galaxies seem to be larger than those in the core regions. Such a result could be explained on the ram-pressure model if the stripping of the gas component of the discs of spiral galaxies occurred at an early phase of the development of the original spiral. Star formation proceeds more slowly near the edges of spiral galaxies and hence the most evolved galaxies would therefore be of larger apparent size. This interpretation could be affected by another important factor which affects galaxy evolution in clusters - 'tidal stripping' - which we shall discuss below.

If the preponderance of E and SO galaxies in rich, centrally condensed clusters like A1656, A426 and A2065 is due to ram pressure stripping, the proportion of E + SO galaxies to S galaxies should increase with time. In this case, distant clusters of similar morphological type should possess smaller (E + SO)/S ratios. This interpretation could be complicated by other evolutionary factors at extreme distances. Recall that we noted that is is probable that most clusters formed at a redshift of about $z = 5$. The young galaxies in the infant clusters would be expected to have a blue colour due to their relatively large population of hot giant stars, which would progressively die out as they rushed through their short but spectacular lifetimes; more being born from the ashes of supernove explosions, but in ever decreasing numbers. Unfortunately, however, galaxies cannot yet be observed at such distances unless they contain quasar hearts. The most distant galaxy cluster with measured redshifts is known as Cl.$13^h05 + 29^o.52$ which has a $z = 0.947$. Recently Butcher and Oemler[10] have studied distant clusters morphologically similar to Coma, with a view to determining the (E + SO)/S ratios. They showed that between 1/3 and 1/2 of the galaxies in Cl.$00^h24 + 16^o54$ and the cluster around 3C295 have the colours of spirals. The latter two clusters lie at redshifts of 0.39 and 0.46, and are therefore observed roughly 8×10^9 years earlier in evolution than A1565.

As a cluster ages, gravitational encounters between orbiting galaxies tend to produce a state of equipartition of energy. The more massive galaxies will tend to move more slowly than the less massive ones, and sink slowly towards the cluster centre, where they move in orbits of relatively small mean radius. The smaller galaxies gain velocity and tend to be boosted to higher orbits. This type of dynamical evolution towards a symmetric equilibrium state is called 'relaxation' and a cluster is fully relaxed when all trace of its initial conditions has been erased. If all the galaxies could be treated as pointlike particles of equal mass, one could treat the cluster as a gas of galactic molecules with a well defined temperature in equilibrium under its gravitational self-attraction - a so-called isothermal gas sphere. Early work by Zwicky demonstrated that the radial galaxy distribution in A1656 closely resembles that expected for an isothermal gas sphere. As we noted above, if a cluster were to attain a condition of equipartition of energy amongst its member galaxies, one would expect a radial mass segregation with more massive galaxies lying near the cluster centre. Since the luminosity of a galaxy generally increases with mass, significant radial luminosity segregation could be used as evidence for equipartition. Amongst the many investigations into this effect is a detailed analysis of 12 rich Abell clusters by Dressler(19) who found evidence for segregation in the clusters A274, A2256 and A2218.

Tidal stripping is an extremely important effect in rich clusters of galaxies where gravitational encounters will be frequent relative to the field. Galaxies are believed to be surrounded by large distended halos of stars, and in a rich cluster, orbital motions through the densely populated core are expected to strip galaxies of their halos and outermost stellar population. Massive, relatively static galaxies at the core will tend to capture these stars and hence grow at the expense of most of the other cluster galaxies. This phenomenon is thought to account for the existence of extremely large galaxies - the cD galaxies - in the cores of many rich clusters. Deep photometric analysis of the radial luminosity profiles of many of these galactic giants reveal halos up to a megaparsec in diameter, filling most of the cluster cores. For example Oemler(12) used microdensitometer photometry down to a surface brightness of 29 mag./sec^{-2} to measure a diameter of about 1 Mpc for the halo of the cD galaxy in the cluster A2670. The most recent work on this topic is due to D. Carter(13) who studied giant galaxies in A2197 and A2199 in Hercules. He showed that the giant 14 mag. galaxy NGC.6166 at the centre of A2199 could be traced to a diameter of at least 700 Kpc, and that NGC6173 in A2197 could be traced to a diameter of at least 600 Kpc. If cD galaxies are the results of tidal stripping (or galactic cannibalism) one would expect the radial luminosity distribution in cD clusters to be effected. Computer simulations show that a bright normal central elliptical galaxy can grow to resemble a cD galaxy with a bright core and a large halo. The dwarf galaxies found close to the cores of many giant galaxies, such NGC.6166 and NGC.4874, may be the undigested, tightly bound cores of galaxies which ventured too close to the giant. A similar observation has been made of a relatively large number of cluster galaxies in the vicinity of the cD galaxy in A2029.

One might expect two types of systematic galactic alignment properties in rich clusters of galaxies. Suppose for example that a rotating protocluster were to collapse into a discoid. One would expect that turbulent motions within the discoid would produce local density enhancements, later to collapse into protogalaxies, with spins parallel or antiparallel to the spin of the whole discoid. In this case there might be a systematic preferential PA alignment for galaxies within a cluster. Such alignments were searched for by L.A. Thompson(14) in eight rich clusters - Virgo, A119, A400, A1656, A2151, A2197 and A2199. It was concluded that a systematic alignment of galaxies occurs in A2197, and that the E galaxies in A1656 appear to be radially oriented with respect to the cluster centre. However, other earlier work on alignments in A151, A1377, A1589, A1930, A2048 and A2065 failed to show any systematic alignments. Thompson's result on A2197 showed a preferential alignment at 90° to a line from the galaxy centre to the cluster centre. The second type of systematic alignment that might be expected is a systematic alignment of an elongated cluster with an elongated giant galaxy - especially in cD clusters. The photographic appearance of cD galaxies is remarkably uniform - (almost always a relatively large oval) which dwarfs nearby cluster galaxies - having an axis ratio of about 1.5:1 - (E3). The alignment of cD clusters with the cD galaxy has been pointed out by many workers. For example, the clusters A2029, A401, A2199, A1413, A2218 and A2670 are known to be aligned. The latter cD is circular and so is the cluster.

As we noted in chapter 2, Abell assumed a degree of universality of the luminosity functions of galaxies in clusters. Such a universality might be expected if the crystallisation of protoclusters followed from a statistically defined spectrum of protogalaxies, and essentially 'froze' after galaxy formation. That this might not be the case is clear from our discussion above of the physical processes which can affect the evolution of clusters. For example, tidal stripping will affect the luminosity function, depleting the number of bright galaxies and greatly increasing the luminosity of first ranked galaxies. Ram pressure stripping will also affect the luminosity function. However there are many indications that there is a useful degree of universality in the luminosity function, and in the following section we shall discuss models of the luminosity functions for rich clusters. After this, the radial distribution of galaxies in rich clusters will be discussed. Both distributions have characteristics which provide useful cosmological distance indications.

3. The Luminosity Function of a Rich Cluster

The differential luminosity function (f_c) of a cluster 'c' is defined in terms of the number of galaxies $dN_c(m)$ in the magnitude range m to m + dm:- $dN_c(m) = f_c(m)dm$. The integrated luminosity function $N_c(m)$ is the number of cluster galaxies with magnitude less or equal to m;

that is, the integral of f_c from magnitude $-\infty$ to m. Clearly, for large magnitudes, $N_c(m)$ will approach a limit which is defined as the total cluster population N_c. Knowing the distance of a cluster, one can convert the apparent magnitude to absolute magnitude by translating along the magnitude axis. The rough form of both functions follows from the protocluster collapse model we have been discussing and these are depicted below:

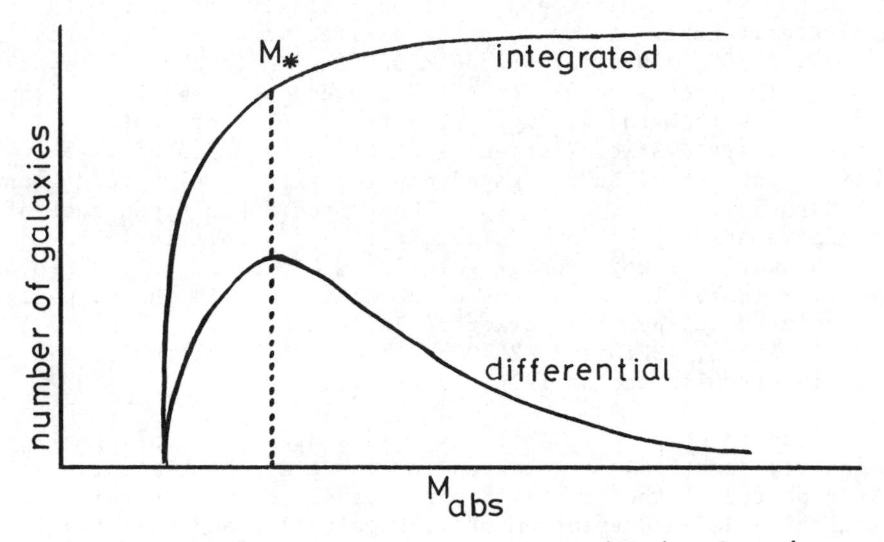

Fig. 2 Schematic form of cluster luminosity functions

Galaxy luminosity is an increasing function of galactic mass. The luminosity function will therefore depend on the mass function for a cluster which, if the mass distribution 'freezes' out after galaxies have been formed, it will resemble the initial mass distribution for protogalaxies which condense out of the protocluster. One would expect relatively few massive protogalaxies to form, but the numbers should increase rapidly as a 'favoured' galaxy mass was approached, subsequently falling slowly for progressively smaller galactic masses which would form the bulk of the objects. One would also expect an upper limit to the mass of a protogalaxy which would be stable to fragmentation. The function might be expected to be of the same for for all clusters, being only dependent on total population, and based upon well-defined dynamics of protocluster collapse. In such a case, the luminosity function for clusters is called 'universal', all cluster luminosity functions being derived from a standard formula either by rescaling the vertical counting axis or translating along the magnitude axis.

If the luminosity functions for rich clusters were universal, they would contain a multitide of parameters which would be useful indicators of cluster distances and richness. Indeed, we have already noted that universality was used by Abell in his analysis of his catalogue of rich clusters. Abell used m_{10} as a distance indicator, and the small spread in the magnitude of first ranked cluster galaxies is a classical cosmological distance probe. Another useful indicator would be the magnitude m_* (or 'turn-over' magnitude) defined in terms of the maximally populated magnitude. In practice, the luminosity functions of clusters are determined from wide field plates of clusters using either fully or semi-automated methods for determining individual galactic magnitudes. Because of the finite size of galaxy images on those plates, the magnitudes are usually defined as the integrated magnitudes out to a given level of surface brightness. The curves are laborious to construct and have to be corrected for many factors such as the contamination by field galaxies and galactic absorption in the Milky Way. Two series of very detailed surveys of a large number of clusters have recently been compiled. These are due to Godwin et al.(15) and to Dressler(11). Godwin et al. constructed the luminosity functions for the clusters A1413, A1930, A1656, A426, A1377, A1553, A2065, A2147, A2151, A2199, A2670 and Virgo, whilst Dressler constructed the luminosity functions for A2256, A2029, A274, A168, A154, A2670, A98, A1940, A1413, A665, A2218 and A401.

Although various analytic expressions have been suggested for a universal luminosity function, the simplest form is due to Abell who suggested that the integrated luminosity function for many clusters could be fitted quite well by an expression of the form:-

$$\log_{10}(N_c(m)) = am + b$$

where the slope a is about 0.8 for $m < m_*$ and about 0.25 for $m > m_*$, m_* being the maximally populated magnitude. This function is represented by a pair of straight lines in the $\log_{10}(N(m))$: m plane, which intersect at m_*. Both of the above mentioned series of results show reasonable agreement with this form. For example, if the magnitudes m_* of the change in slope of the integrated luminosity functions are obtained by an 'eyeball fit' of Abell functions, Godwin et al.'s results yield an average m_* of 20.85 for V(25) - the integrated visual magnitudes out to the 25 mag. are $\sec^{-2}$ isophote. The spread in $V(25)_*$ is from 20.7 to 21.2 - 0.5 mag. if the result for A426, which seems rather low at 20.2, is neglected. Since the cluster redshifts range from z = 0.018 for A426 to z = 0.1851 for A1553, the small spread is remarkable.

4. The Radial Distribution of Galaxies in Rich Clusters

The geometrical shape of many clusters of galaxies such as A2151 (Hercules) or A262 is ill-defined and irregular, with many sub-condensations. Some clusters, however, have a roughly spherical shape, with galaxies crowding together in the central regions. Examples of such clusters include A1656 (Coma) and A2065 (Corona Borealis). Zwicky was the first to investigate the distribution of galaxies in clusters, and his attention was focused on the rich, compact relatively close, Coma cluster. In the absence of tidal interactions, one can model a cluster of galaxies as a constant temperature gas in dynamic equilibrium under gravitational self attraction - a so-called isothermal gas sphere. Zwicky was able to obtain a satisfactory model of the galactic distribution in Coma in this way. Coma is rather a special case in this respect, being one of the most symmetric clusters known. One might expect that most clusters would lack symmetry since only dynamically relaxed systems would have erased all trace of initial subclustering. Indeed, the above mentioned indications of 'missing mass' in many clusters would seem to be good evidence for a general non-equilibrium situation in rich clusters.

Apart from the isothermal gas sphere model for relaxed clusters, many other radial distribution functions for clusters have been suggested. The multitude of current theoretical radial distribution curves suggests the difficulty in determining such distributions observationally. For example, there is confusion due to field galaxies, differing magnitude limited samples and in some cases a lack of a well defined 'cluster centre'. At present, one of the most popular and successful models is the King model, originally formulated to describe stellar distributions in globular clusters and elliptical galaxies, where it is very successful. The latter objects are much easier to investigate, being much better defined, and orders of magnitude richer than the richest cluster of galaxies. The model predicts the following radial distribution of galaxies per square degree of sky to a given magnitude limit:-

$$\sigma(r) = \sigma(o)/(1 + (r/r_o)^2)$$

Here, r is the angular distance from the cluster centre in a symmetric cluster. The parameter r_o is known as the 'core' radius and is defined by the property: $\sigma(r_o) = \sigma(o)/2$, that is, r_o is the radial distance where the surface density of galaxies falls to half the central density $\sigma(o)$.

Many investigators have drawn up radial distribution plots for clusters in order to investigate the possible use of 'core radius' as a cosmological distance indicator. Note that such a parameter is easier to determine than, say, the cluster radius. The latter is an ill-defined quantity in many ways. According to a strict interpretation of models such as the isothermal gas-sphere model or the King model, the distribution should melt away to zero only at infinity. In any case, once the number density falls to about 10% above the density of field galaxies, the determination of radial distribution functions becomes an almost impossible task in practice. One of the most recent investigations of galaxy distribution within clusters is due to Dressler[11], who showed that in a sample of 12 rich clusters, the King model is a remarkably good fit in 10 out of the 12 cases. However, the model fails to predict the remarkably high central concentration of galaxies in the remaining two clusters. These are A2029 and A154. The former is a cD cluster and it is possible that the central excess of galaxies over the King model could be due to the influence of the cD galaxy. According to Dressler's results, the core radius, corrected for linear decrease in size with redshift, is remarkably constant, being of the order of 0.5 Mpc. Similar results have been obtained by other authors, but with varying estimates of absolute core radius.

Recently, the distribution of galaxies in clusters has been investigated using the methods of general relativity theory. Roughly, a cluster is treated as a density perturbation in a background expanding universe. These models are of great interest due to the claims of several authors to have found local minima in the radial distribution functions of several clusters. That is, the galaxy surface density appears to fall initially from the core density to a local minimum value, and then rises and falls again[16]. For example, Clark claimed that A2199 shows a local minimum, whilst Oemler detected minima in 15 clusters using an automated plate scanning method, the effect being strongest in cD clusters. The general relativistic analysis mentioned above does predict oscillations in the radial distribution function, but so far, has not yielded a good fit to observations. Again the interest of the method is that it could yield yet another cosmological distance indicator. However, the method would be difficult to apply in practice due to the same difficulties applying to the core radius, being confused by such factors as sub-clustering. It is unlikely that it will prove as successful a tool as some of the luminosity indicators.

5. Classification Systems for Rich Clusters

Apart from Abell's richness classification scheme for rich clusters of galaxies, there are two main classifications in the Rood-Sastry scheme, and the Bautz Morgan scheme. The former system is based upon the cluster morphology, whilst the latter is a description of the relative dominance of the first ranked cluster galaxy.

(a) The Rood-Sastry Scheme

The Rood-Sastry classification is based upon the distribution of the 10 brightest galaxies within a cluster and can be arranged in a Hubble-type tuning fork diagram:-

```
          L - F
cD - B <
          C - I
```

The various subclasses are described as follows:-

cD. - Class coinciding with Bautz-Morgan class I (see below) in which a cluster is dominated by a single giant galaxy, often with a gigantic halo, perhaps a megaparsec in diameter. On photographs, a cD galaxy will be at least twice the apparent size of the next brightest galaxy, and very often an oval object, about E3 on the Hubble scheme, with a dense halo of dwarf galaxies relatively close by.
Examples:- A2199, cD = NGC.6166; A2634 cD = NGC.7720.

B. - The cluster is dominated by a pair of relatively close giant galaxies much exceeding other cluster galaxies in brightness and size. The binary pair will almost always consist of E galaxies with a swarm of nearby dwarf companions.
Type example:- A1656, B = NGC.4889 and NGC.4874.

C. - The cluster has a core containing at least 3 or 4 of the ten brightest cluster galaxies which, on plates, appear to dwarf other cluster galaxies.
Examples:- A119, A2065 (Corona Borealis Cluster).

L. - At least 3 of the 10 brightest cluster galaxies are distributed in a line or chain upon the sky.
Examples:- A426 (Perseus Cluster) with a chain containing about two dozen galaxies stretching from NGC.1275 to IC310. A194 with a chain of about a dozen bright galaxies stretching from NGC.545/547 to NGC.530.

F. - Some of the 10 first ranked galaxies are distributed in a F = flattened distribution.
Type example:- A397.

I. - Irregular - the 10 brightest galaxies are distributed at random over the cluster which has no well defined symmetry in the galaxy distribution.
Examples:- Virgo, A2151 (the Hercules Cluster), A1367.

The above classification system, although useful, is somewhat ill-defined and subjective. For example when does a line diverge from a flattened configuration? The classes are insufficiently exclusive: the cluster A2197 when examined on a wide field plate might be classed as L, F or even I, whilst one of its brightest galaxies, NGC.6173, is classed as a cD galaxy. As in the case of the Bautz-Morgan scheme, the Rood Sastry scheme is open to contamination by bright foreground galaxies.

(b) The Bautz-Morgan Scheme

As mentioned above, the Bautz-Morgan classification system is based upon the relative dominance of its first ranked galaxy over other cluster members. One can base the system on a range of apparent magnitude differences but, in essence, it can be described as follows.

Class I. The cluster is dominated by a single supergiant cD galaxy as in Rood Sastry class cD.

Class II. The brightest galaxies are intermediate in appearance between cD galaxies (with giant halos) and ordinary giant E galaxies. The Coma Cluster A1656 can be placed in this class since its dominant galaxies, NGC.4874 and 4889, possess well developed halos exceeding those of normal E galaxies in size but being insufficiently large to qualify them as cD galaxies. As another example, one can take A119 in which two of the brightest core galaxies have well developed halos on the POSS.

Class III. The cluster contains no dominant galaxies, i.e:- no galaxy dwarfs all others in brightness and apparent size. Examples:- A1367, A2151.

The Bautz-Morgan system contains a few ambiguous points as formulated above. For example, the distinction between supergiant E and cD galaxies depends upon many factors - one's plate material, exposure time, and whether or not one has traced density profiles on a microdensitometer. A class II cluster might contain a giant E which is a cD, but which does not dominate plates of the cluster because of the presence of other giant E galaxies. That is, the presence of a cD galaxy in a cluster need not imply a BM-I classification. Other objections arise from possible contamination by foreground galaxies. Also, the usefulness of the classification decreases with distance, when the halos of giant E galaxies become more difficult to detect at vast distances, as well as being selectively dimmed due to their intrinsically reddish colour.

Many other cluster classification schemes have been proposed but the two described above, together with the Abell scheme, are the only ones in general usage. In all, the Rood-Sastry scheme seems to have more content, drawing on more factors dependent upon cluster dynamics. It is also easier to apply to very distant clusters, as it is based more upon geometric rather than photographic criteria.

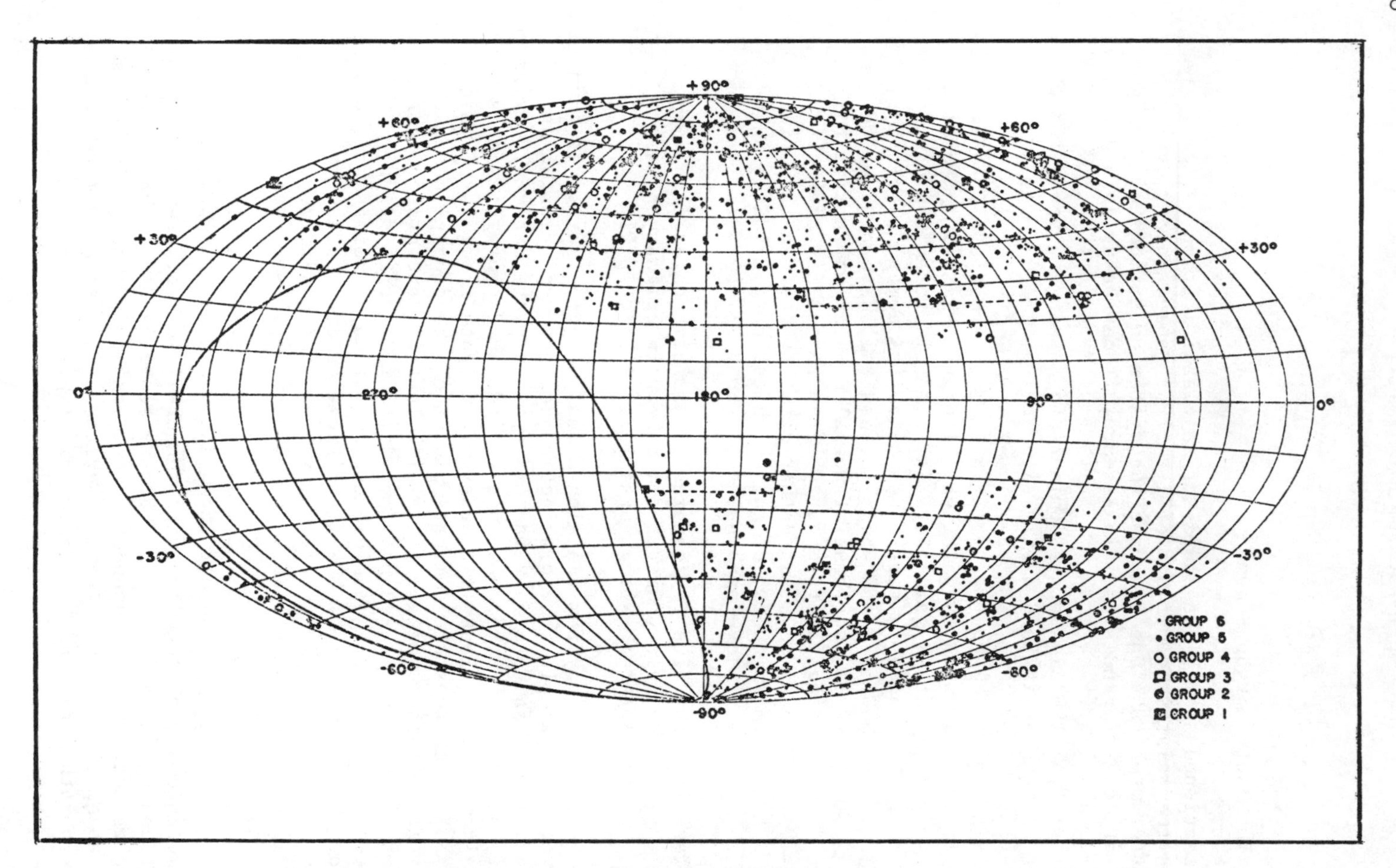

Fig 2A. The Distribution of Abell Clusters in Distance Groups 1 to 6
(Reproduced by kind permission of Prof. G. O. Abell)

4. THE VISUAL OBSERVATION OF CLUSTERS OF GALAXIES

INTRODUCTION

Because most, if not all, galaxies are members of groups and clusters, the techniques required for cluster observation represent only a small extension of the techniques for 'field' galaxy observation. (Covered in volume 4.) However, most of the interesting clusters, such as the nearby Abell clusters, lie at the limit of visual observation with amateur apertures, and push the observer to the limits of his instrument, sky conditions and observing skill. Of course, there is a lifetime's pleasure to be had by observers equipped with almost any telescopic aid at all in observing galaxies in the Virgo cluster and the other nearer small groups. For example, some of the Messier objects in the Virgo cluster can be picked up in a 60 mm refractor whilst 6 inch reflectors show many more. Most of the nearby groups such as the M81 group, the NGC.1023 group and the Local Group are of such a large apparent size that their galaxies are not noticably clustered at all, and these galaxy systems are treated as field galaxies in Volume 4.

As mentioned in the introduction, for the purposes of this volume, a group (poor cluster) or cluster of galaxies will be defined as a concentrated collection of galaxies, either known to be a physical association (e.g. Virgo) spread over a possibly large area of sky, or as an apparent knot, compact enough to yield at least 3 or 4 galaxies in a LP eyepiece with a field of about 30' arc (e.g. the NGC.3158 group). The main extension of visual galaxy observation to cluster observation is in identification, which can become extremely difficult in the case of rich clusters (e.g. Coma A1656) or even require historical research when the authoritative sources of identification are ambiguous or simply wrong. In the following sections, the main techniques required for visual observation of clusters of galaxies are reviewed. Emphasis is placed upon the proper recording of one's observations, this is of great importance to the already difficult problem of identification.

OBSERVATIONAL TECHNIQUE

1. Finding

As in all branches of visual astronomy, finding is the necessary, sometimes difficult and frustrating prelude to observation. Of course, one has to have some idea of the positions of the objects of interest. In the case of clusters of galaxies, it is hoped that the catalogue section of this volume will be a useful guide in this respect. Any serious deep-sky observer should equip him or herself with a good catalogue, and a set of star maps. Although elementary charts like

Norton's Star Atlas may be of use to the novice, more enthusiastic observers will rapidly outgrow them, and probably progress to the Atlas Coeli and its accompanying catalogue. Again, although the atlas does mark some quite difficult objects, its main appeal will be to the owners of small aperture telescopes (8" or less). Its main defects are its paucity of reference stars required by those unfortunates, (like the author) not equipped with setting circles, in the technique of 'star hopping' to one's astronomical quarry. Perhaps the best combination of atlas plus catalogue for the amateur deep sky observer is the SAO atlas plus the Revised New General Catalogue (RNGC) compiled by Sulentic and Tifft. The advantage of the SAO atlas is its high magnitude limit (9.0) and its plotting of all the NGC and IC objects. Unfortunately, the two are slightly incompatible because the RNGC lists positions for epoch 1975 whilst the SAO atlas marks positions for epoch 1950. However, this is not a serious disadvantage for cluster observation where relative positions are most important. The absolute positions quoted by the RNGC translate easily to the SAO by eye or by precession of the 1975 positions back to 1950 using tables or pocket calculator.

There are star atlases with comparable stellar detail to the SAO atlas (such as Atlas Borealis, Eclipticalis etc.) but these do not mark in the nebulae. For the serious, well equipped amateur, the catalogues he requires coincide with those required by the professional. The main reference catalogues are the Reference Catalogue of De Vaucoulers, the Morphological General Catalogue of Vorontsov-Velyaminov, the Catalogue of Galaxies and Clusters of Galaxies compiled by Zwicky, and the Uppsala Catalogue. These catalogues are expensive and difficult of access, and are supplemented by numerous research papers containing individual systems of nomenclature used in the professional investigation of clusters of galaxies.

There are two main techniques for finding a deep sky object. The most convenient way is to look up its position and then to use accurately calibrated setting circles with a well aligned equatorial mounting. Many amateurs are not so equipped, and thus have to use one of the many variants of the 'star-hopping' technique. This process requires a well aligned finder telescope. Opinions vary on the 'ideal' version of this indispensible aid, almost as important as the main telescope it is intended to serve. A wide field of at least one degree is necessary, as well as sufficient light gathering power to show up at least as many stars in its field as there are on one's charts. The latter point can be overstressed, some authors favouring an aperture roughly 1/4 to 1/3 of the main telescope. One can go too far since too many reference stars can be a nuisance. Perhaps a short focus 3" refractor would be an upper useful limit on telescopes up to about 24" aperture, almost all smaller telescopes - 6" to 16" being more than adequately served by a short focus 2" refractor.

The star hopping technique proceeds as follows. Having marked the position of the object of interest on the charts, one locates the nearest visible star with the finder, and compares the field with charts. If the object lies within the field, it is then a simple matter to centre the cross wires of the finder at the approximate position, and then to pick up the object in the main telescope using a low power eyepiece (if it is visible at LP). If LP fails to reveal the object, MP (which may increase the contrast) should be tried. If this fails (as it often does), one has possibly misinterpreted one's charts. The latter possibility will be enhanced if an inverting elbow prism is used in conjunction with one's finder - one has to train oneself in mental inversion of geometric patterns! The ideal here is to use a non-inverting Amici prism. If the visual reference star does not lie in the field, one of several techniques can be used.

Star hopping in its strict interpretation requires a sequence of overlapping finder fields to be constructed, containing recognisable asterims in the overlaps from the visual star to the object. A variant is to locate the object with reference to the geometric layout of visible reference stars on an elementary star atlas (Atlas Coeli is ideal) and then to aim one's finder at the estimated location whilst attempting to pick up a recognisable field around the object. This technique is often necessary as some of the most interesting objects lie in stellar deserts, for example the Virgo Cluster lies beyond such a desert, making finding particularly frustrating. In this case, it is only too easy to pick up a galaxy - any galaxy except the particular galaxy of interest. To aid the technique of star hopping, one should have a very wide field eyepiece allied to ones finder as well as an eyepiece with a somewhat narrower field. Two finders, say a 30 mm and a 60 mm would seem to be near the ideal, the former being used at a power of about 6X, the latter with a power of about 12X. Even a 30 mm can provide too rich a stellar field in some applications, producing too many stars making identification from say, Atlas Coeli, difficult. One very useful device here is a peepsight or unit power finder aligned with one's main finders.

2. Visual Observation

Having found the object, what is actually seen is a function of one's telescope, sky conditions, the object's altitude above the horizon, ones observational ability, and of course, most importantly, the nature of the object itself. These parameters will be discussed under separate headings.

(a) Instrumentation

It is a general principle in visual deep-sky work that the larger one's instrument, the more one sees when comparing instruments of the same type. The latter point is important since telescopes of the same aperture with differing optical systems will have greatly differing performance when used with a given eyepiece. In general, instruments of short focal length are to be the preferred over long focal length instruments. This is because a long focal length instrument (such as a refractor or Cassegrain type reflector) will produce images of a higher magnification in a given eyepiece than a short focal length instrument such as a typical Newtonian reflector, usually with focal ratio between f/4 and f/8. Because of this, nebular images may be so spread out as to be completely invisible. The main difficulty here is in finding an eyepiece of sufficiently long focal length to give low image scales whilst still producing an exit pupil near to the size of one's dilated pupil - implying that the whole field will be seen simultaneously. Again, refractors and catadioptric telescopes contain more surfaces to reduce the precious light available to the observer's eye. A mirror in poor condition may reflect only about 60% of incident light - two such reflections results in only 36% of incident light reaching the eyepiece - a loss of about 1½ magnitudes. Such losses increase with the number of reflections. Similar comments apply to reflection and transmission losses in refractors and in eyepieces. Almost the ideal instrument for any deep-sky visual work is a simple Newtonian reflector, perhaps used in conjunction with a single element Plossol type of eyepiece. The focal ratio should be ideally about f/6, where one obtains the benefit of relatively short focal length without many of the difficulties associated with faster ratios.

The choice of eyepieces is also obviously rather important. All observers should possess at least two good eyepieces. One of long focal length, yielding a power of about 4-7X per inch of aperture, and a wide field, together with a medium power eyepiece yielding about 10-15X per inch of aperture, the choice of focal length depending upon the focal length of the main instrument. The main use of the wide field eyepiece is in finding objects which may then be observed with higher powers, as well as in the observation of objects of large apparent size. For an undriven telescope, LP and MP eyepieces will probably suffice, since the use of HP magnifies the apparent motion of the sky to a degree which makes its use tedious in the extreme. However, the use of HP can sometimes bring out details in individual objects not visible under lower magnifications because of contrast effects. Apart from this type of advantage, HP has little use in deep-sky astronomy, where contrast is more important than image scale as a general rule. Many other details of instrumentation can have an effect upon deep sky work. For example sheer comfort while observing can improve the quality of one's work.

However, details of an 'ideal' mounting and accessories are in the realm of personal preferences and will not be dealt with here.

(b) Sky Conditions Etc.

Contrast is the most important parameter affecting the visibility. of a nebular object. One may define the parameter in the following convenient way:

Contrast = (Surface Brightness Gradient at Sky) × (Mean Surface Brightness Relative to the Sky)

A typical nebular object will have the following type of radial surface brightness variation (solid line).

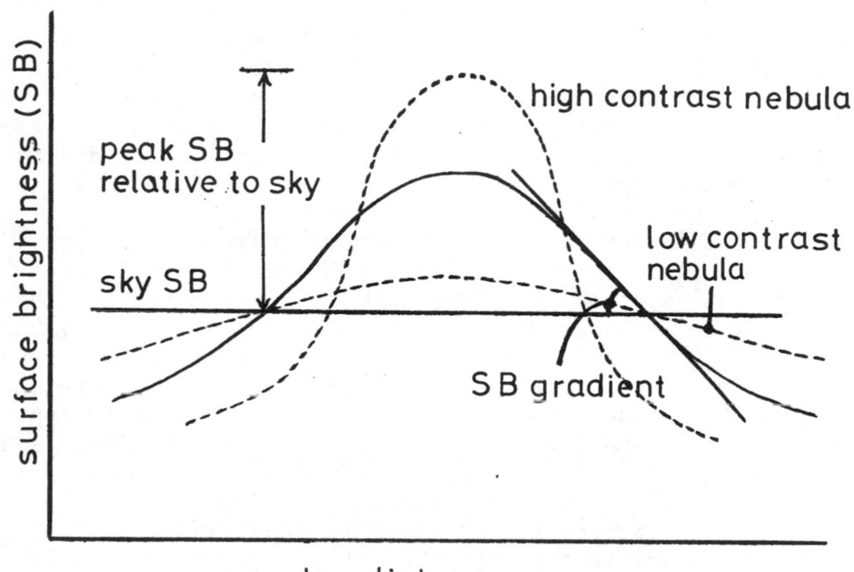

Fig. 3 Surface brightness profiles

The surface brightness gradient at the background surface brightness of the sky determines the crispness of the periphery of the nebula, while the mean surface brightness of the nebula relative to the sky, determines the amount of light energy available to differentiate the nebula against its background. The two quantities are related to the total intensity relative to the sky and hence to effective magnitude.

The contrast parameter is a more useful guide to visibility than the integrated magnitude (or total intensity) commonly quoted by amateurs as visibility guides. For example, an object like M33 has a integrated magnitude of about 6, whilst it is difficult to pick up unless sky conditions are perfect. The explanation is that both its surface brightness gradient at the sky, and mean surface brightness relative to the sky, are at best low. On a really transparent night, the latter parameter is raised, since the sky's average surface brightness level falls. The average surface brightness of the sky is dependent upon local lighting and upon seasonal and atmospheric conditions. Near urban lighting, even a transparent atmosphere will scatter a lot of light, whilst the slightest traces of mist or high clouds can scatter distant lighting, or even more subtle forms of light 'pollution', to raise the average sky brightness. One must also include the effects of direct lighting at the observing site, which can scatter into the eye, and cause an apparent increase in sky brightness.

We next consider the effect of choice of eyepiece on nebular contrast. Most experienced observers will be aware of the fact that contrast is generally best using a MP eyepiece. The explanation is obvious from an examination of the contrast parameter. In short, whilst a long focal length, wide field eyepiece will yield the brightest images due to the relative concentration of available light into a small scale image, the amount of sky background illumination is higher. The latter will sometimes overcome the former advantage and may also reduce the apparent surface brightness gradient at the sky. A HP eyepiece will cover a smaller apparent area of sky yielding a darker background, but the image scale may become so large as to reduce the apparent surface brightness relative to the sky to render an object unobservable. The use of MP represents a compromise - relatively small, dark fields with relatively bright images. Of course, the above comment about HP is less applicable to the observation of small bright objects since then the image scale may still be acceptable. The brightness gradient of a nebula is intrinsic to the object, but it will change along its profile. Thus as the level of sky brightness is varied, the contrast parameter will vary through both the brightness gradient at the sky and the mean surface brightness relative to the sky. For a given object, the observer can only improve the contrast by minimising the effect of sky brightness. Of course, the most effective way of minimising the effect is, (short of obtaining a personal extra-terrestial observing platform) to select a site as far as possible from towns or villages. One should observe an object at or near the meridian (when there is minimal atmospheric extinction and scattering) if possible after midnight when, with luck, most urban lighting is switched off and atmospheric conditions have stabilised. Nebular observation is almost worthless if the moon is above the horizon, or if there is any trace of atmospheric obscuration - one should test transparancy on a well-known test nebula e.g. M31. The effects of direct lighting can be minimised either by using a rubber eyepiece cup or, better still, by covering one's head and the drawtube assembly with a dark cloth. This can sometimes have good results even on the darkest nights.

As far as atmospheric conditions go, transparancy is the most important parameter for the deep sky observer. The other parameter "seeing' is not nearly as important except that it can reduce brightness gradients. Seeing is most important to the observer seeking fine detail or for the astrophotographer. The quality of one's optical surfaces can affect visibility because poorly figured surfaces or ill-designed spiders etc. can degrade contrast by diffraction and scattering effects.

Bright field stars can affect the visibility of nebulae quite markedly. For example the bright field galaxy NGC.404 is situated very close to the 2nd magnitude star beta-Andromeda. Although NGC.404 is intrinsically a relatively high contrast object, the effect of beta-Andromeda is to swamp almost completely an otherwise superb telescopic object. Indeed, this 12 mag. object is not even marked in the Atlas Coeli. As examples of this effect in clusters of faint galaxies, one has the case of A1656 (the Coma Cluster), which is concentrated around an asterism of 6 mag. stars which are a particular nuisance (although making A1656 the easiest cluster to locate) and A1377 (Ursa Major I) whose 15 mag. galaxies are almost swamped by the star BD+56 1544. Whenever possible, such stellar interlopers should obviously be kept out of the field of observation.

Finally, a related parameter is the level of dark adaptation of the observer's eye, the ultimate sensor behind even the most sophisticated visual telescope. Each observer is limited by his or her personal visual activity. However, one's eye is at its most sensitive only after a relatively long period of dark adaptation, and of course, deep sky work should never be attempted unless one has had at least 15 minutes of complete darkness. Avid television viewers should also wait until the dark 'afterburnt' image of the screen on their retinas has had time to fade against the sky. The preservation of dark adaptation whilst recording one's observations is best accomplished by the use of red lighting of variable brightness set on the dimmest usable level. Some observers tape their impressions, although this can never contain as much information as a good field drawing. The old adage that a picture is worth a thousand words is eminantly applicable to the recording of one's observations - particularly observations of crowded clusters of galaxies.

There are various techniques which one can employ to detect faint objects at the limit of vision. Because the eye is most sensitive to very low light levels off its optical axis, the use of an averted vision will often reveal extremely faint nebulae (or details in brighter nebulae) which are invisible with direct vision. Another technique which is sometimes useful is to tap the telescope gently if one suspects a faint object at some point of the field of view. This technique relies upon the eyes' ability to detect subtle changes in surface brightness varying in time. It is sometimes not appreciated that the eye can become trained to be able to pick up faint nebulae. For example after a long period away from the telescope, one can become unaccustomed to this task, although it can be relatively rapidly re-acquired. Another useful approach is to return frequently to an interesting field. At first observations, only the brightest objects or most obvious details will be noted. But once these details have become safely stored in ones memory, they become even more obvious, and one searches for more subtle detail. Two examples of this are in the case of the neublous knots in M33 and, in larger apertures, the swarms of faint galaxies in the Coma Cluster.

(c) Nebular Type

We have seen that the contrast parameter is a good indicator of the visibility of a nebular object. Of course, the visibility of a nebula is dependent upon external factors which affect the apparent sky brightness, but each object will have its intrinsic surface brightness profile defining its visibility in given conditions. The surface brightness profiles of galaxies are functions of Hubble class and orientation in the sky. Typical profiles are depicted schematically below:

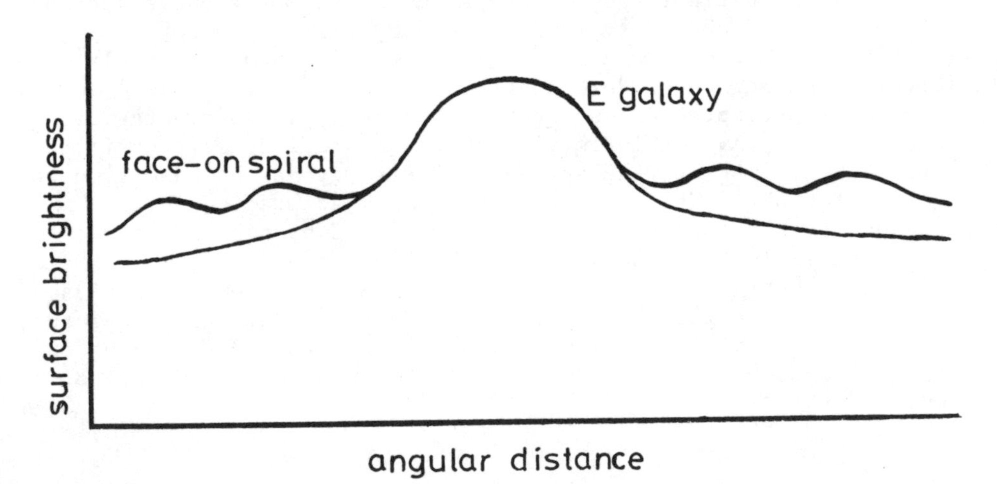

Fig. 4 Surface brightness profiles as functions of galaxy type.

The surface brightness profiles of the E and face-on spiral are similar near the central regions, with a relatively high surface brightness gradient at the sky level. Thus E galaxies and the cores of spiral galaxies will show a comparable high visibility with good contrast. The use of increasingly large apertures will in general raise the level of the brightness profiles, and hence in a sufficiently large telescope the spiral arms of the face-on spiral may become visible. However, on average, surface brightness gradients will be low at the limit of visibility, so contrast will be low compared to the core region. Indeed, for the spiral galaxy, the region of low surface gradients leading to the disc component may well become visible giving a 'soft' texture to the periphery of the disc, whereas an E galaxy may remain of relatively high contrast.

It is easy to see why, in general, edge-on disc galaxies are easier to see than face-on galaxies of similar average surface brightness. For example, the galaxy M74, a face-on Sc spiral is very difficult in small apertures, whereas the galaxy NGC.981, an edge-on system with similarly low surface brightness, is slightly easier. In an edge-on galaxy, the brightness gradients at the sky level are much steeper, yielding a high contrast relative to a face-on system.

When observing clusters of faint galaxies, one is therefore most likely to pick up high contrast systems such as E galaxies, SO galacies and edge-on disc galaxies which have a high contrast at the sky, rather than oblique or face-on spirals with similar integrated magnitudes. A good example here is the Hercules cluster of galaxies (A2151) which contains a high proportion of oblique spirals, and a sprinkling of E and SO galaxies with similar integrated magnitudes. Any visual observation of the cluster will show up the E and SO galaxies before the oblique spirals.

As a final point, the amateur visual observer should be wary of the literal interpretation of photographic magnitude values quoted in catalogues. Integrated visual magnitudes are a little more useful but are still not a complete guide to visibility. The use of integrated magnitudes for nebulae is in some ways a historical relic, and is tied to the method used to measure 'brightness' using wide field, small scale photographic plates, when most galaxies have stellar images. It is of little use in describing the visibility of extended objects, indeed, it is somewhat arbitrary in that most galaxies have large halos which contribute to the true integrated magnitude. That is, their cut-off levels are set by somewhat arbitrary parameters such as emulsion sensitivity and exposure time.

The main use of integrated magnitudes is in providing an upper limit on faintness levels which one may hope to detect. Thus if one's telescope will not show up a 15-th magnitude (high contrast!) star it very probably will not show up a relatively extended object of similar integrated visual magnitude. However, since quoted photographic magnitudes vary, it is sometimes worth trying to pick up very faint galaxies with high quoted magnitudes (within reason) since visual galaxy magnitudes are always higher than (blue) photographic magnitudes. Another factor which helps here is that faint galaxies are usually distant, and hence have a small apparent size and hence may have a reasonably high mean surface brightness.

(d) Galaxy Identification in Clusters

As we have mentioned before, the NGC and IC comprise mainly of visual observations made by a multitude of observers. Most of these observers used rather long focal length instruments which resulted in small fields. Thus in clusters of nebulae, the galaxy identifications quoted in the catalogues are somewhat confused, even the latest revision - the RNGC contains quite a few of these ambiguities. Indeed, once having observed a cluster, the task of deciding just what one has actually observed is an intriguing if somewhat frustrating pastime for those cloudy evenings when the most enthusiastic observer reverts to 'arm chair' status. However, even if the identifications and their accompanying descriptions are taken as gospel, one can still experience great difficulty in identification. Ideally, the observer will be equipped with the CGCG, MCG, RNGC and a complete set of POSS plates, as well as an accurate set of setting circles and so-on. However, most amateurs are not so well equipped and will therefore have to do their best with a subset of these aids. The task of identification will be approached in individual ways varying from observer to observer, but the following system is probably as good as any.

Once a target cluster has been chosen, the observer should first locate the cluster, and survey its area with an LP eyepiece to establish just what is visible. In the case of Abell clusters, the distance class is a useful indicator of how much can be expected to be seen, since it is related to the magnitude of the 10-th brightness galaxy. Having found as many cluster members at LP as possible, these fields should be quickly scrutinised at MP, trying to pick up fainter companion galaxies. One should then, ideally, locate these MP fields in a fairly detailed sketch of the finder field if all the MP fields lie within it, or within an overlapping sequence of finder field sketches. These finder field sketches should all of course have their orientations recorded, and have stars contained in overlapping fields identified in both fields. Such field sketches will prove invaluable when attempting to identify one's main observations from catalogues. Next, one should sketch as much of the cluster as possible, using overlapping LP fields if possible, marking some of the stars in overlapping fields. These sketches should mark in all faint field stars, as one might later find by consulting the catalogues, that they are actually almost stellar galaxies. This accomplished, each galaxy field should be carefully drawn using MP, again overlapping as much as possible, and carefully noting the orientation. A note should also be kept of the main features of the individual galaxies, such as their shape, size, surface brightness gradients, close companions, close field stars etc. as well as the sky conditions.

The drawings should all be made at the eyepiece, using a dimmable red light preferably fixed to the telescope or one's clipboard. The effect of the drawing light (which can also obviously be used for chart reading) is to temporarily, impair one's night vision. To try to avoid the effect of this, it is wise to survey the field thoroughly, trying to pick up all it has to offer before starting the sketch. Once one has an idea where everything is, the sketch can be started, keeping one eye closed whilst drawing if possible, to try to preserve its dark adaptation.

Having completed one's observation, the task of identification follows. One way to simplify one's task is to prepare a chart of the relative layout of the cluster galaxies at roughly half the scale of one's LP observations, using catalogue positions. It is not an uncommon occurrence in cluster observation to pick up anonymous (i.e. non NGC or IC) galaxies. In this case one should seek identification in the CGCG, MCG or Uppsala Catalogue and confirmation from POSS plates. For those lucky enough to have access to library facilities, one can also sometimes obtain off-prints of research papers which contain copies of large aperture plates of clusters, and this is an invaluable help in confirmation of observations, as well as identification. Another source of field plates is textbooks: many volumes contain the occasional photograph of galaxy clusters. Some references will be found in this volume, together with charts of some of the cluster fields. It is important to note that one should reobserve clusters as often as possible, since as one's familiarity with their fields grows, one sees more and more new detail.

Another useful aid is to have access to some chart of the cluster during an observing session. Whilst it might be argued that this might bias the observer into 'imagining' he has observed objects he could not possibly see, it is up to the conscience of the individual to be as objective as possible; indeed, no 'advantage' is to be gained from making outrageous claims. The benefits that are to be gained lie in the fact that knowing exactly where to look is a great help when observing at the limits of vision, and one can hence pick up more. It is perhaps wise not to take the charts to the telescope, but to refer to it just before observation, retaining an approximate position in one's memory. To the author's mind, this is no more cheating than, say, observing a galaxy, checking one's description against the NGC, and finding that there is a very faint nearby companion and subsequently picking it up. One's first sweeps should be made with the charts safely stowed indoors. In this way one gains great satisfaction in comparing one's drawings with the chart. Subsequent use of the charts to pick up more objects cannot detract from this.

PART TWO : CATALOGUE OF CLUSTERS OF GALAXIES

INTRODUCTION

The catalogue contains visual observations of clusters of galaxies made by Webb Society members using telescopes from 5 inch to 82 inch - mainly reflectors - and is split into three sections. Firstly, the Virgo cluster is treated separately because its member galaxies are so bright and relatively isolated due to their closeness. Its observation serves as an introduction to the much more challenging if sometimes less spectacular distant groups and Abell Clusters of galaxies. Also, because of its huge apparent size and richness in bright galaxies, limitations of space have imposed the need to consider only a limited part of the cluster in this volume (the richest part), which will be called the central region, covering an area of sky of about 200 square degrees and containing about 110 galaxies down to the 13th photographic magnitude.

The second part of the catalogue will cover the visual observation of some of the nearer Abell clusters of galaxies. The clusters were chosen from the catalogue on the basis of their Abell distance class, which as we noted earlier, is defined in terms of the apparent photographic magnitude of the 10th brightest galaxy. Recall that distance group 3 corresponds to M_{10} ranging from 14.9 to 15.6 mag., and that distance groups 0 to 2 contain rather brighter galaxies. The lower limit is just about as faint as most amateur visual observers can go using telescopes of up to about 16"-18" aperture. Observers with access to 24" or larger telescopes may be able to pick up galaxies down to magnitude 16 or more, and thus have access to distance group 4 Abell clusters. Of course, for $M_{10} \leq 15.6$, one might still be able to pick up the brightest cluster members which shine at about mag. 14 to 14.5 to register an observation. Some of the observations presented here are of this type.

The third part of the catalogue is concerned with the visual observation of relatively sparsely populated compact groups of galaxies. These are defined as those knots of NGC/IC galaxies which should contain at least 3 galaxies in a MP eyepiece field of about 15' to 20', and to some rather more extended groups which are known to be physical associations. Examples of groups of the first type include 'Stephan's Quintet', an old favourite of both amateur and professional astronomers, and the NGC.383 group, which is sometimes known as the 'Pisces Group'. Examples of the less concentrated or more extended type of group are the Pegasus I cluster and the NGC.5416 group. The groups were mainly chosen by the author from the pages of the SAO, where they are liberally sprinkled over the charts of regions away from the plane of the galaxy. Unfortunately, most contain rather faint galaxies, almost all of them fainter than the 13th photographic magnitude, and they thus require telescopes of modest aperture to begin to show something of their full splendour - say 12 inches and above.

In short, most of the objects in the second and third sections of the catalogue require telescopes of large aperture by amateur standards. The enthusiastic owners of smaller telescopes should, however, find amply rich galaxy fields in the Virgo Cluster.

Each set of rich cluster observations will be prefaced by a short introduction containing (hopefully) interesting physical data (much of it very recent) on the cluster. This will be followed by a finder chart, taken from POSS plates where possible, together with galaxy identifications from catalogues, or with those identifiers currently used in the professional literature. These charts will be followed by a series of field sketches containing cluster galaxies, together with verbal descriptions, all being made by the various observers listed at the beginning of each separate sub-catalogue. Any ambiguities in galaxy identification found in the RNGC for the cluster fields are pointed out and, where possible, solutions are provided from research in the more detailed catalogues and/or by reference to the POSS. In the case of most of the groups finder charts for the individual galaxies are not required because of their compactness.

1. THE VIRGO CLUSTER

1. Description of the Cluster

The Virgo Cluster is the nearest of the rich clusters of galaxies, and is consequently spread over a large area of the sky. The cluster has been subdivided into subclusters by some authors, but we shall apply the name to the whole of the Local Supercluster, which can be traced on the sky from a declination of about -20° in southern Virgo through Virgo, Coma Bernices and Canes Venatici into Ursa Major at a declination of $+55^\circ$. Various estimates of its distance have been obtained by a variety of independent methods, such as the apparent size of the largest HII regions in spiral galaxies or the apparent magnitudes of the brightest globular clusters associated with galaxies. These estimates yield a relatively large range of disagreement, ranging from about 12 Mpc to about 20 Mpc, the latter distance being based upon the mean redshift of galaxies in the cluster:- $\langle V \rangle = 10^3$ km sec^{-1} and $H_o = 50$ km sec^{-1} Mpc^{-1}. It is interesting to note that the velocity dispersion within the cluster is quite large, velocities ranging from -500 km sec^{-1} to 2500 km sec^{-1}. Thus some galaxies have orbital velocities high enough to dominate the small cosmological recession velocity at this close distance, and to show blue-shifted spectra. If we take the figure of 20 Mpc as the distance of the centre of mass of the cluster (situated near the giant E galaxy M87 = NGC.4486 which is almost at rest relative to the centre of mass), the apparent angular size of about 75° noted above yields a figure of about 20 Mpc for its major axis. It is indeed fortunate for deep sky observers that by some lucky accident of nature, our galaxy is oriented almost face-on to the Virgo Cluster which is spread about in the region of the north galactic pole. Otherwise, much of its splendour would be extinguished by clouds of obscuring dust in the plane of the Milky Way.

From now on, for reasons of economy of space, our discussion of the Virgo Cluster will be limited to its central region, which we define as that part of the cluster lying within $12^h.00$ to $13^h.00$ of RA and declinations $+07^\circ00$ to $+20^\circ00$ (1950), most of the brighter galaxies and crowded low power galaxy fields being situated within this region. In fact, many of the brightest galaxies in the whole sky including 13 of the 39 galaxies in the extended Messier catalogue (taking M91 = NGC.4548) are to be found in the central region. Atlas Coeli includes 83 of the galaxies in the region which are brighter than the 13th photographic magnitude, whereas the Shapley Ames catalogue (13 mag. survey) lists 107 galaxies to the 13th magnitude. These are plotted on a large scale finder chart in the following pages, the chart also depicting field stars down to the 9th magnitude as shown on the SAO. In the SAO Atlas, all NGC and IC galaxies in this region are plotted amongst field stars, but here there are so many galaxies that identification is rather difficult. The RNGC lists 215 galaxies in the region (in 1975 coordinates) down to photographic magnitude 15.5, where comparison with large scale POSS plates indicates that the catalogue is probably rather incomplete. Using the RNGC data, one can plot the luminosity functions for the central region of the cluster, obtaining the histograms shown below. (Note that the RNGC magnitudes are 'rounded off' to the nearest half magnitude.)

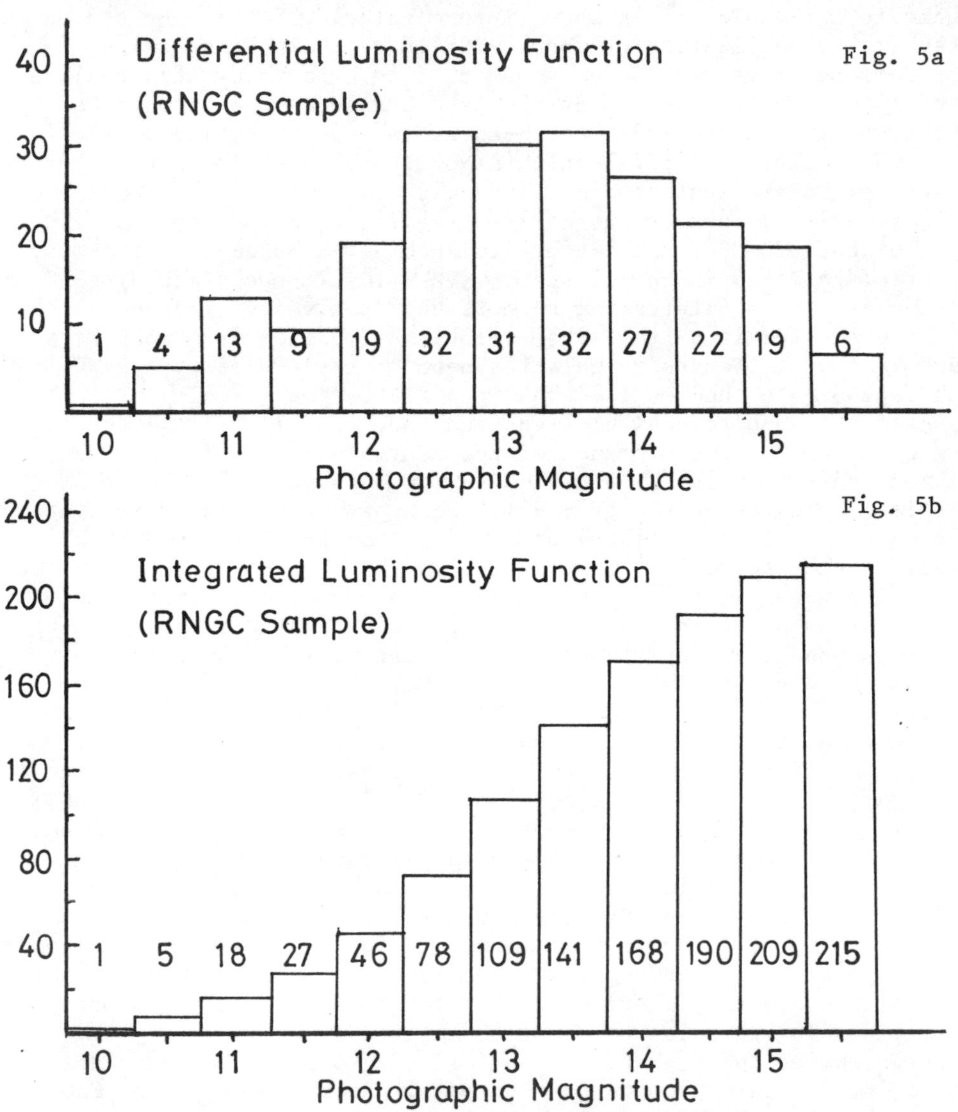

Fig. 5 Luminosity functions for the Virgo Cluster

The above curves are of some practical interest to the visual observer as well as being of theoretical interest. Firstly, the differential luminosity function clearly indicates that the central region is mainly populated by galaxies of apparent magnitude 12.5 to 13.5, with an apparent deficiency of galaxies in the magnitude range 11 to 11.5. A similar deficiency in these galaxies was obtained by Godwin et al. using a sample taken over a different area. Another cluster with a 'wobble' in its luminosity function is the Coma Cluster A1656 which has an apparent deficiency of galaxies in the range $15.5 \leq Mp \leq 16.0$.

The integrated luminosity function (the running total of galaxy counts for the differential luminosity function) is also useful. Noting that $M_3 = 10.5$ and that 73 galaxies lie in the brightness range M_3 to $M_3 + 2$, we obtain an Abell richness class of AR = 1 for the central region of the Virgo Cluster. One can also use the integrated luminosity function as an indicator of how many galaxies one might expect to see in the central region using a given telescope. For example, assuming a magnitude limit of 11.5 for a 6 inch reflector one might expect to pick up 27 of the brightest galaxies (including all the Messier objects). Using a 16 inch with a magnitude limit of about 15.5, one should be able to pick up all the RNGC galaxies, as well as all the IC objects, that is, all roughly 600 galaxies marked in the SAO in this area.

As an illustration of the methods used to calculate such parameters as radial distribution functions and core radius for galaxy clusters, all the NGC + IC galaxies in the survey area were counted in degree squares and plotted as follows.

20	11	8	6	13		3		2	2			1		1	
					1	2	1	1	4	1		2			
							4	4	4		1				18°
		2			1	1	4	5	10	1	1		2		
			1		3	4	7	6	5	2	1	2			
1			1		11	4	5	6	4	5	9	6	6		
		2		5	7	5	8	15	8	5	13	9	5		
		1	2	9	8	5	11	11	13	9	10	8	3	1	
		4	5	12	4	11	13	12	5	6	3	4	2	2	
1			5	6	6	8	1	6	12	4	3	2	4	3	
		1	3	2		1	6	9	10	14	14		1	3	
	4		1	3	4		5	11	7	4	2				8°
				1	2	5	2	2	12	5	5	2			

13h ... 12h

Fig.6 (above) Surface density in galaxies per square degree of NGC+IC galaxies in the central region of the Virgo Cluster.

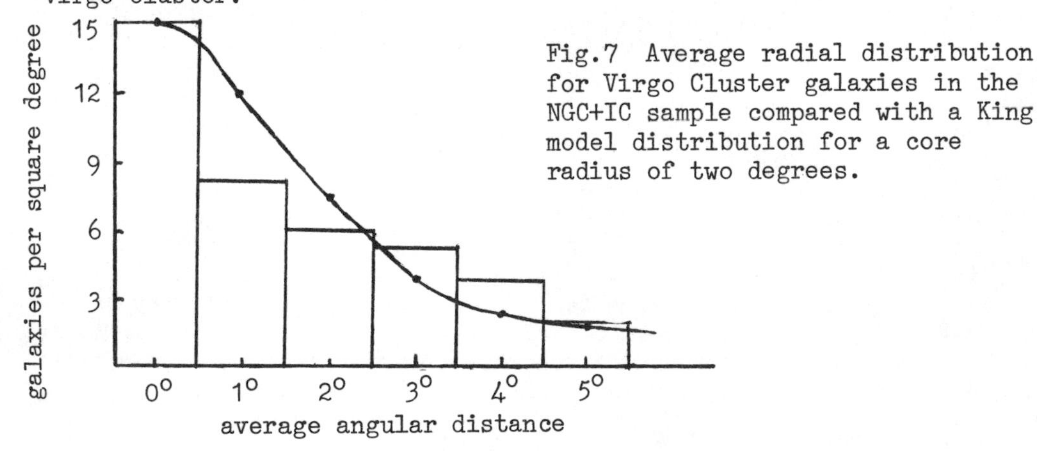

Fig.7 Average radial distribution for Virgo Cluster galaxies in the NGC+IC sample compared with a King model distribution for a core radius of two degrees.

It is readily apparent from this plot that there is extensive sub-clustering in the Virgo Cluster, and that the distribution is very anisotropic. One particular subcluster is to be found in the area at about RA 12^h40-13^h00 and about $+20^o00$, mostly composed of faint IC galaxies. This is very probably a background cluster, and so we shall ignore it in what follows. We shall also ignore the sparsely populated region $12^h48'$-13^h00 $+07^o00$ to 20^o00. The highest surface density of galaxies was found in the square $12^h24'$-$12^h28'$, $+13^o00$ to 14^o00 (near M84 and M86) with a count of 15 galaxies per square degree. This was taken as the cluster centre, although this is more usually taken to be in the south following square containing NGC 4486 = M87. To average out the anisotropy due to sub-clustering, average galaxy densities were calculated in a series of nested squares (each 2° larger in diameter than the preceding square) for 1 degree squares in the area between the inner and outer squares. This is known as the method of 'ring counting'. In this way, the variation of galaxy surface density with angular radius shown in fig. 7 was obtained. In fig. 7 the calculated average radial distribution is compared with a King model curve relating surface density to angular radius θ:-

$$\sigma(\theta) = \sigma(o)/(1 + (\theta/\theta_c)^2) \text{ (note } \sigma(o) = 15)$$

where θ_c is defined as the core-radius: $\sigma(\theta) = \sigma(o)/2$ when $\theta = \theta_c$. The resulting angular core-radius of 2° at 20 Mpc defines a true core radius of about 0.5 Mpc, which compares reasonably well with the universal figure of 0.5 Mpc mentioned in Part One, Chapter 3.

The most studied and most interesting galaxy in the Virgo Cluster is the giant elliptical galaxy NGC.4486 = M87. The galaxy is most famous for its optical jet, an object consisting of several blobs of extremely energetic material emerging from its nucleus and shining through the mechanism of synchroton radiation which is produced by the swirling of very energetic electrons along magnetic field directions. A counterjet has also been detected in the direction opposite to the optical jet. Many theories have been advanced to explain this jet and one possibility is that since M87 is almost at rest at the cluster centre, intracluster matter will tend to be sucked into its nucleus by gravitation possibly producing a cataclysmic explosion. Indeed, M87 is also a strong X-ray and radio source, and it is possible that these radiations are also produced by matter falling into the nucleus. Another piece of evidence to support this is the presence of a large optical halo around M87, probably consisting of tidally captured stars from close encounters with cluster galaxies passing through the core. It has also been proposed that the extremely large number of globular clusters surrounding M87 may be due to the relatively long incubation of the galaxy in this respect, having approximately 5 times as many attendant globular clusters than other average Virgo Cluster galaxies. Only the second most massive galaxy

NGC.4472 = M49 has a similar large number of globular clusters. Recently two independent investigations have pointed to the strong possibility of the presence of a black hole in the nucleus of M87.

Although the Virgo Cluster does not contain any head-tail radio sources, studies of M87 have indicated the presence of an intracluster medium. Other evidence comes from the presence of a large proportion of anaemic spirals near the cluster core. The most pronounced example is NGC.4569 = M90 as pictured in the Hubble Atlas of Galaxies. In the outer part of the galaxy, the spiral arms are completely devoid of the usual bright HII regions and associated clusters of bright, young, massive stars. The only knots to be found lie near the edge of the bulge component, and these are probably due to obscuring dust left behind as ram pressure, due to orbital motion through the intracluster medium, stripped the galaxy of its star-forming interstellar gas. More anaemic objects are to be found deeper into the core. For example the edge on system NGC.4388 in the central region is possibly a stripped spiral, as is its near neighbour NGC.4402.

One would expect galaxy to galaxy tidal encounters to be frequent in rich clusters of galaxies such as Virgo. Evidence of a halo of tidally stripped stars surrounding M87 has already been mentioned. The most spectacular instance of tidal interaction in the Virgo Cluster is given by the pair NGC.4435 and 4438. NGC.4435 is an E or an SO galaxy of magnitude 12 close NP 4438 which is difficult to type because of the existence of two spectacular opposed tidal filaments emerging from the ends of its major axis. These extend for tens of kiloparsecs, presumably having been drawn out by a close encounter with NGC.4435. Recently, deep photography with the UK 48 inch Schmidt at Siding Spring has detected the presence of an optical jet emanating from the 11th magnitude elliptical NGC.4552 = M89. The jet is a much less spectacular one than that in M87, and is currently supposed to have been caused by an ancient tidal encounter with a long since separated galaxy. Another example of a possible pair of galaxies in tidal interaction is the pair of overlapping 12th mag. spiral galaxies NGC.4567/4568. Readily available photographic material does not reveal any evidence of significant distortions or tidal effects, so this pair may be a line of sight effect.

2. Visual Observation of the Virgo Cluster

In the Virgo Cluster it is easy to pick up galaxies with almost any telescope. Indeed, one of the most pleasurable of deep sky pursuits is to sweep at random across it, picking up this or that object, and then effortlessly sweeping on to other splendours. However, systematic observation is rendered more difficult in these crowded regions than in the general field. One really needs a plan of campaign, which in the relatively sparse stellar foreground field, needs to be well thought out, if one is not to become hopelessly lost. Most methods, such as that suggested by Kenneth Glyn Jones[1] and by Roland S. Copeland[2], rely upon a combination of star and galaxy-hopping. One such route, which is as good as any, is to start near the 6 mag. star rho-Virgonis which is marked on chart (1) of the 13 mag. galaxies. Owners of telescopes of 8" aperture or larger will dwell a while on NGC.4608 and 4596, preceding rho by about ½°. Sweeping north by a little over a degree, one will come across the 10.5 mag. galaxy NGC.4649 = M60, a giant E galaxy easily visible in a 60 mm finder. Its close attendant Sc galaxy NGC.4647, which is shown in contact on large aperture plates appears a few minutes of arc NP visually. NGC.4647 is of mag. 12, and should be an easy object for most apertures. The binary system is known as Arp 116, appearing in H.C. Arp's collection of peculiar galaxies. Following M60 by about 25', one comes across NGC.4621 = M59, an 11.5 mag. elliptical galaxy.

There are several other galaxies in this area worthy of observation. One of these is NGC.4638, a 12.5 mag. object with a tiny spindle-shaped companion close following. The latter is numbered NGC.4637 in the RNGC, although it follows 4638, and is rather difficult visually, although it has a quoted magnitude of 12.5. The visual difficulty was also probably apparent to its visual discoverer - hence the mix up in numbering in the NGC.

North preceding M59 by about 4' of RA and about 20' of arc, is another Messier object, the 11th mag. SB spiral NGC.4579 = M58. This interesting object can be used as a reference point for sweeps to the north and south. About 2' of RA preceding, and about 45' N, lies the 11.5 mag. elliptical galaxy NGC.4552 = M89, rather undistinguished in comparison with other Virgo cluster E galaxies. About 1.5' of RA following, and about 40' N, lies the bright anaemic spiral, NGC.4569 = M90, with apparent photographic magnitude 11.0. Returning to M59 via M89, one finds the pair of overlapping spirals, NGC.4567/4568 (named the 'Siamese Twins' by Copeland) about 40' S and 1.5' of RA preceding M59. This spectacular object is easy in a 10 inch, and should also be visible with an 8 inch, shining at about the 12th mag. The pair is probably the brightest pair of galaxies which overlap visually, resembling a butterfly's wings.

To continue one's review of the cluster, one can sweep north to pick up once again M89, after which, sweeping westwards by a little over a degree, one arrives at the magnificent galaxy NGC.4886 = M87, near the cluster's centre. This is certainly the most massive cluster galaxy, but according to the RNGC it is outshone by NGC.4472 (M49), another galactic giant E galaxy, by mag. 10 to mag. 11. Depending upon one's telescopic power, one might pick up one or two faint companion galaxies - NGC.4476 and 4478 lie close, SP, shining at about Mp = 13.5.

and 12.5. From M87, it is but a short hop of about 5' RA to the west and 20' N to the beginning of 'Markarian's Chain' - the masterpiece offered by the cluster. This presents a spectacular linear arrangement of about a dozen bright galaxies starting near NGC.4406 = M86, an 11 mag. E3 (or possible SO) galaxy with a large halo, to NGC.4501 = M88, a beautiful oblique spiral. At low power, the field of M86 is a truely awe-inspiring sight, and should contain at least four or five galaxies in a field about 30' of arc in diameter in a 6 or 8 inch telescope. The field will be dominated by M86 and NGC.4374 = M84, an SO galaxy with apparent photographic magnitude of 11.0. A somewhat fainter galaxy forms the vertex of an equilateral triangle with base vertices M84 and M86 - this is an edge-on peculiar system, NGC.4388.

Following M86 by less than 2' of RA is a close pair of galaxies aptly named 'The Eyes' by Copeland. These are NGC.4435 and 4438, with magnitudes of 12.0 and 11.0 respectively. Slightly NF is NGC.4461 and 4458, another close pair separated by about 3 or 4 minutes of arc. 4461 is the brighter spindle-shaped object (12.5 mag.), and 4458 is a roundish object of magnitude 13.5. Slightly to the N again, and slightly following, is a wider pair of galaxies, NGC.4473 and 4477. The former is an elongated object (an elliptical or an SO) of magnitude 12, while NF is NGC.4477 an SB galaxy of magnitude 11. Users of larger telescopes will also pick up NGC.4479, SF 4477 shining at mag. 14. Moving slightly north, yet another superb galaxy swims into view, NGC.4459, a 12th magnitude elliptical with a star close SF. Slightly to the north, and following by about 15' of arc, is still another pair of galaxies - NGC.4474 a 12 mag. spindle, with a roundish faint companion preceding by about 6 or 7 minutes of arc. The latter galaxy NGC.4468 is of mag. 14, and will require at least a 12 or 14 inch for detection. The last object in the chain is the large oblique spiral galaxy NGC.4501 or M88 at mag. 11.0.

Many other rich galaxy fields exist in the central regions of the Virgo Cluster, but none, perhaps, as facinating as Markarian's chain - truly one of the most superb deep sky spectacles imaginable. Having described the most obvious delights, one can adapt the galaxy-hopping method to step off to explore the rest of the cluster, moving north, say, to the fields of M100 (NGC.4321) and M85 (NGC.4382) or south to the fields of M61 (NGC.4303), M49 (NGC.4472) and so on. Of course, the Virgo Cluster extends well beyond the boundaries which have been delineated in this volume. For example, one of its most magnificant objects, the 'Sombrero' galaxy (M104 or NGC.4594) lies a good distance to the south at a declination of over -11°. This and other bright Virgo Cluster galaxies were covered in volume 4 of this series.

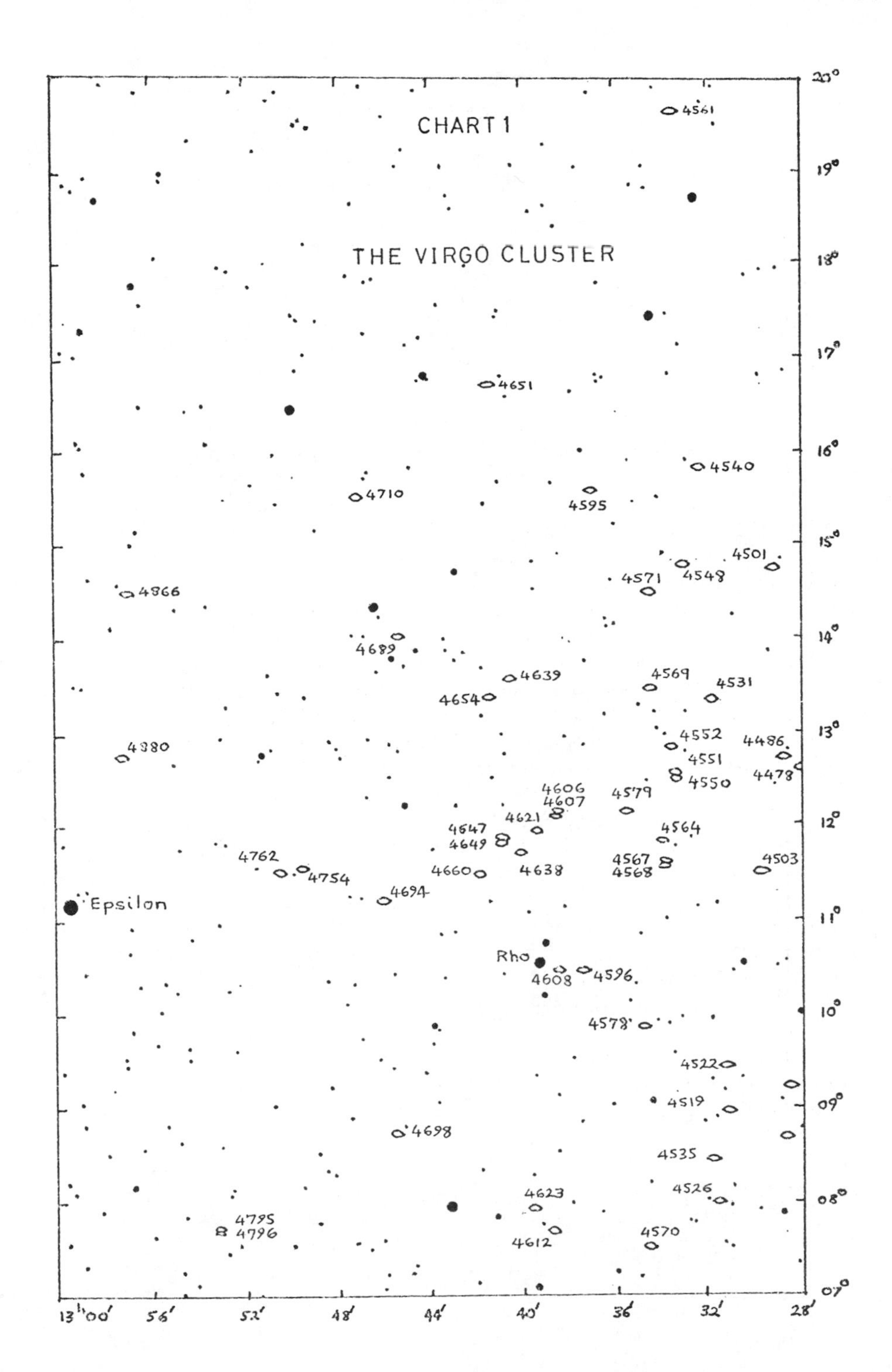

CHART 1
THE VIRGO CLUSTER
4561
4651
4540
4710
4595
4501
4571
4548
4866
4689
4639
4569
4531
4654
4552
4486
4880
4551
4478
4550
4606
4607
4579
4621
4647
4649
4564
4762
4754
4660
4638
4567
4568
4503
4694
Epsilon
Rho
4608
4596
4578
4522
4519
4698
4535
4526
4623
4795
4796
4570
4612
20°
19°
18°
17°
16°
15°
14°
13°
12°
11°
10°
09°
08°
07°
13h00′
56′
52′
48′
44′
40′
36′
32′
28′

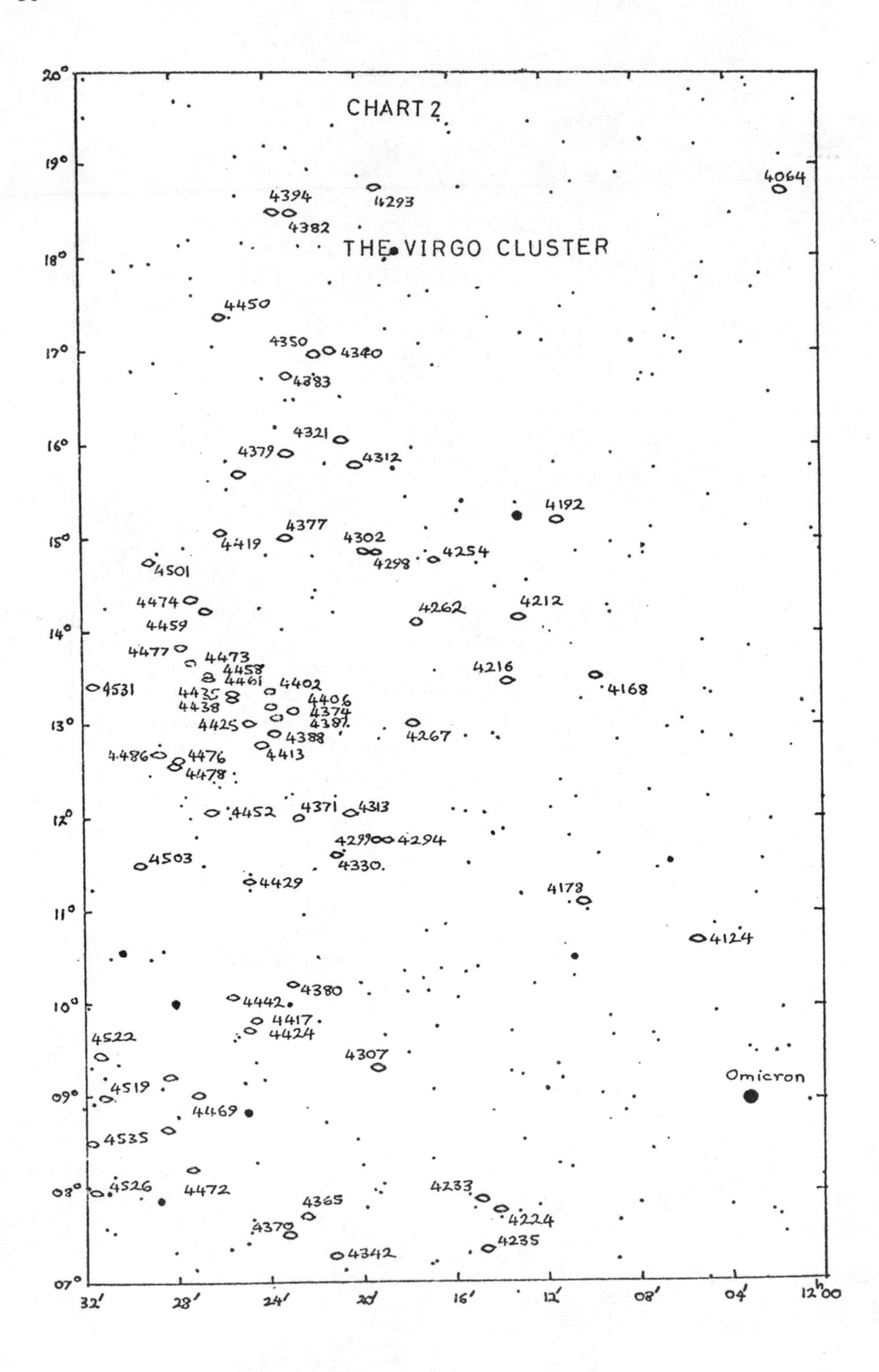
CHART 2
THE VIRGO CLUSTER
20°
19°
18°
17°
16°
15°
14°
13°
12°
11°
10°
09°
08°
07°
32′
28′
24′
20′
16′
12′
08′
04′
12h00
4064
4394
4293
4382
4450
4350
4340
4383
4321
4379
4312
4192
4377
4419
4302
4298
4254
4501
4474
4459
4262
4212
4477
4473
4458
4461
4216
4168
4531
4435
4438
4402
4406
4374
4387
4425
4388
4267
4486
4476
4478
4413
4452
4371
4313
4299
4294
4503
4330
4429
4178
4124
4380
4442
4417
4424
4522
4307
Omicron
4519
4469
4535
4233
4526
4472
4365
4224
4370
4235
4342

(a) List of Observers and Instruments

The following list shows the names of the observers whose work on the Virgo Cluster appears in the following pages, plus details of their locations and respective telescopes.

D.A. Allen	60"	Mt Wilson, California USA
D.A. Allen	12" OG	Cambridge University UK
D. Branchett	5"	Eastleigh, Hampshire UK
D. Branchett	12 × 80 20 × 50 }	binoculars
D. Fielding	11 cm	Salford, UK
K. Glyn Jones	8"	Winkfield, UK
P. Maloney	11 cm	Salford, UK
R.J. Morales	10"	Arizona, USA
J. Perkins	10"	Kirby-in-Ashfield, UK
K. Sturdy	6"	Helmsley, UK
M.J. Thomson	16½"	Santa Barbara, USA
J.B. Thorpe	18"	Salford, UK
G.S. Whiston	16"	Witley, Surrey, UK

The following table lists the galaxies marked on charts (1) and (2) of the central regions of the Virgo Cluster. Note that it is quite possible that some of these objects are in fact either under-luminous foreground galaxies or very luminous background galaxies. Indeed, in the case of M86, Kenneth Glyn Jones[(1)] quotes Holmberg's distance modulus which appears to place this galaxy in the foreground whilst Sandage states (in the Hubble Atlas) that M86 is a cluster member. This problem is characteristic of all clusters of galaxies in the absence of distance indicators independent of assumptions of absolute luminosity and redshift. The latter is most troublesome for the special case of the Virgo cluster because orbital radial velocities can dominate the Hubble velocity. In the case of M86 and all the galaxies on the charts, the probability that M86 is, in fact, in the cluster core, is high and we shall assume therefore, that all the galaxies listed are cluster members.

Included with the list are the names of those observers who submitted observations on the particular galaxy for the Webb Society files. These observations are then summarised in the later pages. Following the verbal descriptions is a set of drawings of cluster galaxies made at the telescope. Most of them were made using a 16" f/5 Newtonian, but almost all of the galaxies depicted are easy objects is 8" telescopes, quite a few (mostly the Messier objects) being visible in a 60 mm refractor or a 4" reflector.

(b) List of Galaxies on Charts with Observers

Galaxy NGC	M	Observers	Galaxy NGC	M	Observers
4064			4419		
4124			4424		DB
4168			4425		RJM
4178			4429		GW,DB
4192	98	GW,KGJ,DB	4435		GW,JP,DB,RJM
4214		JP,DB	4438		GW,JP,DB,RJM
4216		GW,KGJ,DB	4442		GW,DB
4224		RJM	4450		DB
4233		RJM	4458		GW,RJM
4235		DB,RJM	4459		GW,DB
4254	99	GW,KGJ,KS	4461		GW,DB,RJM
4262			4469		
4267		DB	4472		GW,KS,DB
4293		GW,DAA	4473		GW,RJM
4294		GW,JP	4474		GW,DB
4298		GW,DB	4476		GW,KS
4299		GW	4477		GW,DB,RJM
4302		GW	4478		GW,K,JP,DB
4307			4486	87	GW,KGJ,KS,JP,JT,PM,DB
4312		GW,DB,RJM	4501	88	GW,KGJ,KS,DB
4313		GW	4503		GW,DB
4321	100	GW,KGJ,JBI,KS,DAA,RJM,DB	4519		MJT,DAA,DB
4330			4522		MJT
4340		GW,DB	4526		GW,MJT,KGJ,DAA,DB
4342			4535		MJT,DB
4350		GW,DAA,DB	4540		
4365		GW,DB	4548		KGJ,DAA,DB,
4370			4550		GW,MJT,KS,DB
4371		GW,JP,DB	4551		GW,MJT,KS
4374	84	GW,KGJ,DB	4552	89	GW,KGJ,DB,MJT
4377			4564		GW
4379			4567		GW,MJT,DB
4380		DB	4568		GW,MJT,DB
4382	85	GW,KGJ,DP,DAA,DB	4596	90	GW,DB
4383			4570		GW,DB
4387		GW,RJM	4571		DAA
4388		GW,JP,KGJ,DB,RJM	4578		DB
4394		GW,JP,DAA,DB	4579	58	GW,KGJ
4402		GW,RJM	4595		
4406	86	GW,KGJ,JP,DB	4596		GW,KGJ,DAA,DB
4413		DB,RJM	4606		GW
4417		DB	4607		GW

Galaxy		Observers	Galaxy		Observers
NGC	M		NGC	M	
4608		GW,DB	4654		DB
4612		GW,DB	4660		
4621	59	GW,KGJ,KS,JP,DB	4689		DB
4623			4694		
4637		JP,DB	4698		DB
4638		GW,JP	4710		
4639			4754		GW,KGJ,DB
4647		GW,KGJ,KS,JP,DB	4762		GW,KGJ,DB
4649	60	GW,KGJ,KS,JP,DB	4795		
4651			4796		
			4866		

The following catalogue contains verbal descriptions of about 70 of the galaxies on the charts made by Webb Society visual observers using telescopes from 60 inches to 60 mm. Prefacing these descriptions is a list of data on each galaxy:- RA and Dec (1975), photographic magnitude and Hubble type.

(c) Catalogue of Visual Observations

Cat.	RA	Dec.	Mp	Type	Comments
NGC.4192 M98	$12^h12.6'$	$15^o03'$	11.0	SB	Almost edge on.

(16) Bright and large, very elongated. Much brighter to the middle to an oval nuclear region which seemed aligned at about 45° to the main axis. The galaxy seemed curved rather like an integral sign, especially at the NP end.

(8) Quite large, fairly diffuse. Very elongated in PA 150° and slightly brighter to the central area.

(15x80) Faint, large oval.

Cat.	RA	Dec.	Mp	Type	Comments
NGC.4212	$12^h14.4$	$14^o03'$	12.5	S	

(10) Exceptionally faint. Round and brighter to the middle. An orange star follows by 16'.

(15x80) Faint, small elliptical.

Cat.	RA	Dec.	Mp	Type	Comments
NGC.4216	$12^h14.6$	$13^o71'$	11.0	Sb	Almost edge on.

(16) Very bright, very large and much extended roughly N/S. Suddenly much brighter to the middle to a bright almost stellar nucleus. About 8' × 1'.

(8) A long streaky nebula. Very bright in the centre with a slightly rounded nucleus. Something of a dark streak visible. Clearly aligned in PA 20°. Bright star 15' NF mag. 8.9, bright star mag. 9.5 10' preceding.

(5) Bright, large streak, stellar centre.

Cat.	RA	Dec.	Mp	Type	Comments
NGC.4224	$12^h15.3'$	$07^o36'$	13	S	Edge on.

(10) Largest of 3 in a group (4224, 4233, 4235). Average brightness patchy appearance, smaller than average.

Cat.	RA	Dec.	Mp	Type	Comments
NGC.4233	$12^h15.9'$	$07^o46'$	13.0		

(10) Smallest and faintest of group of 3 nebulae.

Cat.	RA	Dec.	Mp	Type	Comments
NGC.4235	$12^h15.9$	$07^o20'$	12.5	S	Edge on.

(10) S of 4224 and 4233. Edge on but irregular shape, patchy appearance. Similar brightness to 4224.

Cat.	RA	Dec.	Mp	Type	Comments
NGC.4254 M99	$12^h17.6$	$14^o34'$	10.5	Sc	'The Pinwheel'

(16) Very bright, large with diffuse halo suddenly much brighter to an almost atellar nucleus. Overall round nebula with clear spiral structure. The southern arm can be traced for about 90° whilst the northern is apparent due to a small relatively dark notch oriented EW.

(8) Quite large, about 5' in diameter. Bright central condensation. with glimpses of several small brightish patches on S & F border.

Cat.	RA	Dec.	Mp	Type	Comments
NGC.4293	12^h200	$18^o32'$	11.5	Sa	Pec.

(16) Bright and slowly brighter to the middle. Quite large, about 5' × 1' and extended approx. EW. Group of stars due NF all about mag. 12 or less.

(12) Elongated streak, about 6' × 1.5', of uniform surface brightness.

Cat.	RA	Dec.	Mp	Type	Comments
NGC.4294	$12^h20.1$	$11^o39'$	13.0	S	Pec.

(16) Preceding 4299 by about 8'. Elongated with PA approx. 45°. Quite bright and slowly brighter to middle. Overall low surface brightness.

(10) Very faint, invisible at 56X and 96X, just visible at 132X. Elongated EW. No other detail noted.

Cat.	RA	Dec.	Mp	Type	Comments
NGC.4298	$12^h20.3$	$14^o45'$	12.5	?	

(16) Close P 4302 (4'). Round, quite bright but low surface brightness. Diameter about 2'. Star of mag. 11 or 12 close NF.

(5) Very faint, vague opaque object. Small.

Cat.	RA	Dec.	Mp	Type	Comments
NGC.4299	$12^h20.4$	$11^o39'$	13.5	?	

(16) Following 4294. Quite a bright oval with diameter about 2'. Slowly brighter to the middle.

Cat.	RA	Dec.	Mp	Type	Comments
NGC.4302	$12^h20.5$	$14^o45'$	12.5	S	Edge on.

(16) Elongated cigar oriented approx. N/S. Close following 4298. Quite bright and slowly brighter to the middle but low surface brightness. Very faint star on N tip suspected.

Cat.	RA	Dec.	Mp	Type	Comments
NGC.4312	$12^h21.3'$	15^o41	13.0	?	

(16) Faint with uniform low surface brightness. Quite large cigar shaped nebula (about 5' × 1') oriented approx. N/S.

(10) Much extended, slightly brighter to the middle. No definite core.

(5) Faint small streak, opaque.

Cat.	RA	Dec.	Mp	Type	Comments
NGC.4313	$12^h21.4'$	$11^o57'$	13.0		

(16) Quite bright and much extended with PA of about 45°. Faint stellar nucleus. Possible absorption lane to S of this nucleus. Slowly brighter to the middle (3.5-4' × 45").

Cat.	RA	Dec.	Mp	Type	Comments
NGC.4321 M100	$12^h21.7'$	$15^o58'$	10.5	Sc	

(18) Small, brighter than expected. Nucleus elongated N/S. Bright star involved.

(16) Quite large and round with a diameter of about 5'. Diffuse halo suddenly much brighter to a brilliant nucleus. Some traces of spirality but not as clear as in M99.

(12) A 5' diameter circular nebula. Quite faint and then much brighter to the middle.

(8) Circular distinct patch with a much brighter almost stellar nucleus and a mottled or brush-like perimeter.

(6) Nebula about 4'-5' × 2' with a stellar nucleus.

(5) Bright, large oval, bright nucleus.

Cat.	RA	Dec.	Mp	Type	Comments
NGC.4340	$12^h22.4'$	$16^o52'$	12.5	SB	

(16) Round, about 1' in diameter. Quite bright and much brighter to the middle. NP 4350 by about 8'.

(5) Bright elliptical, brighter to the centre.

Cat.	RA	Dec.	Mp	Type	Comments
NGC.4350	$12^h22.7$	$16^o50'$	12.0	E	

(16) Spindle with PA of about 320°. Size about 2' × 30". Quite bright and brighter to the middle.

(5) Bright lenticular in field of 4340.

Cat.	RA	Dec.	Mp	Type	Comments
NGC.4365	$12^h23.2$	$07^o28'$	11.0	E	

(16) Bright oval, about 2' × 1.5'. Diffuse but slowly brighter to the middle. PA of about 330°.

(5) Bright elliptical, very opaque.

Cat.	RA	Dec.	Mp	Type	Comments
NGC.4371	$12^h23.7$	11^o51	12.0	SBO	

(16) Bright oval, about 2' × 1', PA about 70°. Brighter to the middle with a small core containing a faint stellar nucleus. No sign of dark lane mentioned in RNGC.

(10) Round and bright. MP shows the galaxy well and a star like nucleus visible. Mottled. Appeared as bright as M84 situated close north.

(5) Bright large oval, brighter to the centre, very opaque.

Cat.	RA	Dec.	Mp	Type	Comments
NGC.4374	$12^h23.8'$	$13^o02'$	11.0	E	

(16) Very bright and much brighter to the middle with a brilliant core. Round and about 2' in diameter.

(8) Round. Bright and brighter to the middle. But the nucleus is not well defined.

(5) Bright, large oval, very opaque, becoming brighter to centre.

Cat.	RA	Dec.	Mp	Type	Comments
NGC.4382 M85	$12^h24.2'$	$18^{\circ}20'$	10.5	E	

(16) Very bright and very suddenly much brighter to the middle to a brilliant core. Diffuse halo. Star superposed on N tip of comparable brightness to the core. About 3' × 2'. PA 320°.

(10) Very bright and round. Faint star on northern extremity. NGC.4394 visible in same field.

(8) Very small, about 3' in diameter but bright and concentrated. On some occasions extent grows to about 5'. Small, fairly bright, elongated nucleus.

(5) Bright large oval becoming brighter to the centre, very opaque.

Cat.	RA	Dec.	Mp	Type	Comments
NGC.4387	$12^h24.5$	$12^{\circ}57$	130	E	

(16) Quite small, about 20" major axis. Slightly elongated with uniform high surface brightness.

(10) Round and brighter to the centre. Star close to N edge.

Cat.	RA	Dec.	Mp	Type	Comments
NGC.4388	$12^h24.5$	$12^{\circ}48'$	12.0	Pec.	Edge on.

(16) Quite bright and much brighter to the middle. Elongated approximately E/W. Star close NF central region.

(10) Faint and elusive, needs averted vision. Elongated east-west. HP shows brightening to the centre.

(8) Faint blur 20'S of M84. Only central region visible.

(5) Faint, elliptical, compact, very opaque.

(10) Bright gradually brighter to the middle. Oriented E/W. Longer than M84 or M86.

Cat.	RA	Dec.	Mp	Type	Comments
NGC.4394	$12^h24.7'$	$18^o21'$	12.0	SB	

(16) Quite bright and slowly brighter to the middle. Cigar elongated with PA approximately 45°. About 1' × 30". About 10' following M85.

(12) Small elliptical nebula following M85, oriented at PA approximately 45°. About 1' in diameter.

(10) Almost stellar even at HP.

(5) Faint, small, very close to M85.

Cat.	RA	Dec.	Mp	Type	Comments
NGC.4402	$12^h24.9'$	$13^o15'$	13.5	E?	

(16) Quite bright, slightly brighter to the middle. Equatorial lane mentioned in RNGC not observed. About 2' × 30" and oriented approximately E/W. 10' NP M86.

(10) Very elongated, even brightness. Oriented E/W possible brighter E half.

Cat.	RA	Dec.	Mp	Type	Comments
NGC.4406	$12^h25.0$	$13^o05'$	11.0	E	(Possible SO.)

(16) Very bright, large (about 2' × 1.5'), very much brighter to the middle to a brilliant nuclear core. Oriented approximately 45° from N.

(10) Quite conspicuous, a little brighter than M84. Much brighter to the middle.

(8) Similar to, but slightly inferior to M84. Slightly elongated in PA approximately 80°. Brighter to the middle.

(15x80) Bright, large oval.

Cat.	RA	Dec.	Mp	Type	Comments
NGC.4413	$12^h25.3$	$12^o45'$	13.5	SB	

(5) Faint, small elliptical NF two stars.

Cat.	RA	Dec.	Mp	Type	Comments
NGC.4417	$12^h25.6$	$09^o44'$	12.5	E	Spindle.

(10) Round patch with star like centre. Two stars N.

(5) Large bright elliptical.

Cat.	RA	Dec.	Mp	Type	Comments
NGC.4424	$12^h26.0$	$09^o3.41$	12.5	S	

(5) Faint and small. Best seen with averted vision.

Cat.	RA	Dec.	Mp	Type	Comments
NGC.4425	$12^h26.0'$	$12^o53'$	13.0		Spindle

(10) Elongated and gradually brighter to the middle. Faint star preceding.

Cat.	RA	Dec.	Mp	Type	Comments
NGC.4429	$12^h26.2'$	$11^o15'$	11.5	S	

(16) Bright and much brighter to the middle to a stellar nucleus. Spindle elongated in PA of about 70°. About $2.5' \times 45''$.

(5) Very bright, very large streak with a star SF at edge of system.

Cat.	RA	Dec.	Mp	Type	Comments
NGC.4435	$12^h26.4$	$13^o13'$	12.0	E4	

(16) Bright and much brighter to the middle to a non-stellar core. Oval about 1.5' × 45", oriented parallel to NGC.4438 close following which is slightly superior.

(10) Easy nebula following M86 and close to similar galaxy NGC.4438. Brighter to the middle, well defined oval.

(10) Smaller and rounder than 4438 but brighter than 4438. Almost stellar core. Less diffuse overall than 4438.

(8) Bright and easy close NP similar object NGC.4388.

(5) Bright oval, very opaque.

Cat.	RA	Dec.	Mp	Type	Comments
NGC.4438	$12^h26.5$	13^o09	11.0	Pec.	

(16) Similar to but superior in size and brightness to 4435. Much brighter to the middle to an inconspicuous stellar core.

(10) Bright and large. Larger and slightly fainter than 4435. Elliptical shaped with diffuse core.

(10), (8) See NGC.4435.

(5) Bright elliptical, very opaque.

Cat.	RA	Dec.	Mp	Type	Comments
NGC.4442	$12^h26.8'$	$09^o57'$	11.5	SBa	

(10) Very bright spindle with bright central region

Cat.	RA	Dec.	Mp	Type	Comments
NGC.4450	$12^h27.3$	$17^o14'$	11.5	S	

(5) Bright ellipse. Very large, oriented approximately N/S. Very long, very opaque, star SP.

Cat.	RA	Dec.	Mp	Type	Comments
NGC.4458	$12^h27.7'$	$13^o28'$	13.5	E	

(16) Small, about 30" major axis. Oval. Quite bright and slightly brighter to the middle. Close NP 4461.

(10) Elliptical shaped and brighter to the middle. NP 4461.

Cat.	RA	Dec.	Mp	Type	Comments
NGC.4459	$12^h27.8'$	$14^o07'$	12.0	E	

(16) Bright and slowly brighter to the middle which contains a small bright core. Slightly elongated (about 2' × 1.75') with PA of about 70°. Bright star (mag. about 9 F 2').

(5) Bright ellipse with a condensed centre with a soft diffuse halo.

Cat.	RA	Dec.	Mp	Type	Comments
NGC.4461	$12^h27.8'$	$13^o20'$	12.5		

(16) Bright spindle which is much brighter to the middle to a concentrated almost stellar core. Diffuse wings. About 1.5' by 30" with PA of about 90°.

(10) Brighter of two (4458). Bright compact core and possible stellar nucleus.

(5) Bright ellipse.

Cat.	RA	Dec.	Mp	Type	Comments
NGC.4472 M49	$12^h28.5$	$0.8^o09'$	10.0	E1	

(16) Very bright and suddenly much brighter to the middle to an intense core region with a bright nucleus. Quite large, about 2' × 1.75' and oriented approximately N/S. Star of magnitude about 9 or 10 SF outer edge of nebula.

(8) Bright, round glowing patch about 5' in diameter with a clear outline. Round much brighter central area.

(6) Round, about 5' in diameter. Bright with a bright non stellar nucleus embedded in a bright core. Star involved in halo.

(5) Very bright, large oval. Condensed to an almost stellar core.

Cat.	RA	Dec.	Mp	Type	Comments
NGC.4473	$12^h28.6'$	$13^o34'$	12.0	E	

(16) Bright and much brighter to the middle to an almost stellar nucleus. Spindle about 2-3' by 45" with PA approximately E/W.

(10) Bright elliptical core with faint long extensions. Slightly brighter and less diffuse than 4477 in the same field. Suddenly much brighter to middle.

Cat.	RA	Dec.	Mp	Type	Comments
NGC.4474	$12^h28.7'$	14^o13^o	13.0	Sa?	Edge on.

(16) Bright and much brighter to the middle with an almost stellar nucleus. About 2' by 30" with PA roughly E/W. NGC.4468 SP by about 7'.

(5) Faint, small, vague.

Cat.	RA	Dec.	Mp	Type	Comments
NGC.4476	$12^h28.8$	$12^o29'$	13.5	E	

(16) Quite bright and brighter to the middle. Small oval about 20" major axis. 6' NP NGC.4478.

(6) Very faint, 1' in diameter oval. NP 4478.

Cat.	RA	Dec.	Mp	Type	Comments
NGC.4477	$12^h28.8'$	$13^o47'$	11.5	SB	

(16) Bright and slightly elongated (about 2.5' × 2') with PA about N/S. Diffuse halo slowly brighter to the middle to a tiny core with a stellar nucleus. NGC.4479 SF by about 6'.

(10) About the same size as 4473 (in some field), but fainter and more diffuse. Round with a bright compact core, gradually much brighter to the middle.

(5) Bright oval with almost stellar central condensation.

Cat.	RA	Dec.	Mp	Type	Comments
NGC.4478	$12^h29.1$	$12^o28'$	12.9	E	

(16) Bright and very much brighter to the middle. Round, about 1' in diameter.

(5) Faint oval. Close to M87.

Cat.	RA	Dec.	Mp	Type	Comments
NGC.4479	$12^h29.1$	$13^o43'$	14.0	?	

(16) Quite bright and slightly brighter to the middle. Irregularly round. About 30" in diameter. South following NGC.4477.

(10) Similar size to 4477, but much fainter. Round and slightly brighter in the middle.

Cat.	RA	Dec.	Mp	Type	Comments
NGC.4486 M87	$12^h29.6$	$12^o32'$	11.0	EO	

(16) Very bright and very much brighter to the middle and then very much brighter to an intense non-stellar core. Round, about 3' in diameter.

(10) Quite large and round, being quite bright in the middle fading slowly to the edge. NGC.4478 to S.

(8) Very bright and distinct. The brightness falls off gradually from a bright non stellar nucleus. 4478 to S.

(6) About 6' in diameter with a bright non stellar nucleus.

(5) Very bright, very large oval, very opaque.

(60 mm) Bright globular galaxy. Perfectly circular outline. Very much brighter in the centre. Brightness gradient even.

Cat.	RA	Dec.	Mp	Type	Comments
NGC.4501 M88	$12^h30.8'$	14^o34	11	Sc	

(18) Very large and elliptical shaped. Nucleus off centre. Incomplete ring,(very faint) around nucleus. Diffuse.

(16) Bright and brighter to the middle which brightens rapidly to a bright nucleus. Quite large, about 5' × 1.5' with PA approximately 50°. Diffuse halo with a faint star embedded in SF tip.

(8) Bright oval extended in PA of about 135° with a distinct elongated nucleus. With averted vision a considerable amount of faint detail can be made out from time to time.

(6) Nebula 6' × 3' with distinct non stellar nucleus.

(15x80) Bright, large oval.

Cat.	RA	Dec.	Mp	Type	Comments
NGC.4503	$12^h30.9'$	$11^o19'$	12.5	E	

(16) Bright and much brighter to the middle. Spindle about 1.5' × 30" oriented with PA of about -20°.

(5) Bright ellipse, very opaque, lovely object.

Cat.	RA	Dec.	Mp	Type	Comments
NGC.4519	$12^h32.3$	$08^o48'$	12.5	S	

(16½) Pretty bright, large, diffuse, brightening a little in the middle to a small lens. It is irregularly round. A pretty bright star lies SF whilst just off the NP edge is a faint star. At 176X, the surrounding envelope is mottled.

(12) Nebula 2.5' × 1.5'. Irregularly round, no nucleus.

(5) Faint oval, very opaque.

Cat.	RA	Dec.	Mp	Type	Comments
NGC.4522	$12^h32.4$	$09^o19'$	12.5	S	Edge on.

(16½) Pretty bright, obviously elongated SP/NF. It is quite narrow and only slightly brighter in the middle. At 176X it is seen as a long narrow ray being wider on the SP end.

Cat.	RA	Dec.	Mp	Type	Comments
NGC.4526	$12^h32.8$	$07^o51'$	11.0	E	

(16½) Very bright, very large and greatly elongated NP/SF. At low power it shows a stellar nucleus surrounded by an oval lens. Other faint galaxies noted NP.

(16) Very bright and very much brighter to the middle to a stellar nucleus. Spindle very elongated in PA ∿E/W.

(12) Pear shaped nebula, about 4' × 2' with bright elongated nucleus offset to one end.

(8) Very bright, elongated with PA approximately 105°. Size 4' × 1.5'. Bright centre (non stellar) elongated and possibly brighter on the north side.

(5) Faint oval between two stars.

Cat.	RA	Dec.	Mp	Type	Comments
NGC.4535	$12^h33.1$	08^o21	11.0	S	

(16½) Quite bright, large and irregularly round brightening to the centre to a lens with pretty bright nebulosity surrounding. Star invested inside NF edge. Mottled envelope with a dark lane between central lens and northern edge, the mottling and dark lane give strong impression of spirality. Other faint stars involved visible at HP.

(5) Extremely faint, elusive smudge.

Cat.	RA	Dec.	Mp	Type	Comments
NGC.4548 M91	$12^h34.2$	$14^o38'$	11.5	SB	

(12) Oval nebula considerably brighter to the middle. 3' × 2'.

(8) Quite large and bright. Brighter towards the centre but no clear nucleus. Elongated with PA 40° or 50°. 4571 visible in same LP field SF.

(5) Very faint, large, opaque.

Cat.	RA	Dec.	Mp	Type	Comments
NGC.4550	$12^h34.3$	$12^o22'$	12.5		

(16½) First of a double system, whilst a 3rd galaxy lies almost directly N on edge of the field (4552). Bright and elongated N/S. Brighter to the middle to an extended nucleus. A pretty bright star lies SF.

(16) Quite bright spindle elongated N/S. Uniform high surface brightness. Quite small. NGC.4551 NF 6'.

(6) P M89. Oval nebula of uniform surface brightness close NGC.4551.

(15x80) Faint, large streak.

Cat.	RA	Dec.	Mp	Type	Comments
NGC.4551	$12^h34.4$	$12^o24'$	13.0	E	

(16½) 2nd of a double system (with 4550). Quite bright, medium size and oval, lying SP/NF. Brighter in the centre to a well developed lens also oval in shape. No well defined nucleus.

(16) Quite bright and much brighter to the middle. Small oval about 45" elongated approximately E/W.

(6) Very faint, no nucleus.

Cat.	RA	Dec.	Mp	Type	Comments
NGC.4552 M89	$12^h34.4$	$12^o42'$	11.5	E	

(16½) Very bright, round and brighter to a central core within a lens. Star close north following.

(16) Bright, round and much brighter to the middle with a bright nucleus.

(8) Very small, about 3' diameter. Perfectly round having a well marked central condensation which is almost stellar. Bright.

(15x80) Bright, large oval.

Cat.	RA	Dec.	Mp	Type	Comments
NGC.4564	$12^h35.2$	$11^o35'$	12.5	E	

(16) Bright spindle oriented with PA approximately 330°. Much brighter to the middle. About 1.5' × 30". Between two stars mag. 12.

Cat.	RA	Dec.	Mp	Type	Comments
NGC.4567	$12^h35.3$	$11^o24'$	12.5	Sc	Double in contact.
NGC.4568	$12^h35.3$	$11^o23'$	12.0	Sc	

(16½) NGC.4567 is the first of a double system which appears to almost touch. It is quite bright and oval, the major axis being SP/NF almost on the parallel. Either a stellar nucleus or a small star invested eccentric to the F end. NGC.4568 seems slightly brighter and larger than 4567 being spindle shaped SP/NF at a greater angle to the parallel so that the F ends of both objects appear to be in contact. Like two teardrops. 4568 is of uniform surface brightness except for a small central brighting.

(16) Both bright and slowly brighter to the middle. In contact at following ends at about 60°. 4567 oval about 3' × 2' with a small circular core, oriented approximately E/W. NGC.4568 is similar if a little brighter with a large, less well defined ovaloid core. Similar size to 4567.

(5) Bright, very large fan shaped object.

Cat.	RA	Dec.	Mp	Type	Comments
NGC.4569	$12^h35.6$	13°18'	11	Sb	
M90					

(16½) Very bright, very large, elongated SP/NF with a pronounced stellar nucleus and a large central lens. The latter is quite wide whilst the two extensions do not narrow to points. The NF end is narrower than the SP. Resembles M31 at MP.

(16) Very bright, quite large. Diffuse halo becoming much brighter to a relatively small core containing a bright stellar nucleus. About 7' × 2' with PA approximately 30°.

(8) Nebula about 7' × 3' with PA of about 20°. Bright elongated central nucleus. Distinct and easily seen.

(15x80) Bright, large, oval.

Cat.	RA	Dec.	Mp	Type	Comments
NGC.4570	$12^h35.6'$	07°23'	12.0	5?	Edge on.

(5) Bright spindle oriented N/S.

Cat.	RA	Dec.	Mp	Type	Comments
NGC.4578	$12^h36.3$	09°42'	12.0	SO	

(5) Faint oval best seen with averted vision.

Cat.	RA	Dec.	Mp	Type	Comments
NGC.4579 M58	$12^h36.5'$	$11^o58'$	11.0	SBc	

(18") Small, bright, circular with a definite nucleus. Bar and portions of arms visible NF and SP.

(16½) Very bright. Large mottled envelope. The central area contains a bright core in an extended bar of nebulosity which lies SP/NF. The nebulosity appears brighter on the NP side of the nucleus.

(16) Very bright oval of about 2' × 1' elongated approximately E/W. Much brighter central area with a stellar nucleus.

(8) Similar to M60 nearby, but a little longer and brighter with a central condensation. The nucleus appears extended E/W whilst the ends of the minor axis are better determined than those of the major axis which lies E/W.

(20x50) Large bright oval.

Cat.	RA	Dec.	Mp	Type	Comments
NGC.4596	$12^h38.7$	$10^o19'$	12.0	SB	

(16) Bright oval about 1' × 30" oriented approximately EW. Much brighter to the middle with a stellar nucleus. Larger and brighter than 4608 following.

(12) Elongated nebula, slightly brighter in the centre.

(8) Very faint, only slightly brighter in the centre. No stellar nucleus. PA possibly 50°.

(5) Bright oval, very opaque.

Cat.	RA	Dec.	Mp	Type	Comments
NGC.4606	$12^h39.7$	$12^o03'$	12.5	E	

(16) Quite bright spindle much brighter to the muddle to a stellar nucleus. PA approx. -30°. About 1.5' × 30". Star involved near NP tip. Also very faint star involved near NF tip. 4607 F 5'.

Cat.	RA	Dec.	Mp	Type	Comments
NGC.4607	$12^h40.o'$	$12^o02'$	14.5		Edge on.

(16) Quite bright but considerably fainter than 4606. Very elongated approximately N/S. About 2.5' × 20". Uniform low surface brightness.

Cat.	RA	Dec.	Mp	Type	Comments
NGC.4608	$12^h40.0'$	$10^o10'$	12.5	SB	

(16) Quite bright oval, small, major axis about 1', PA about 310°. Much brighter to the middle to a stellar nucleus.

(5) Bright large oval. Very opaque.

Cat.	RA	Dec.	Mp	Type	Comments
NGC.4612	$12^h40.3'$	$07^o27'$	12.5	SBO	

(16) Very bright oval about 30" major axis. Almost uniform high surface brightness. Oriented approximately E/W. Star of magnitude about 10 close SF. Lies at the end of a chain of 4 bright stars.

(5) Bright irregular streak with a star following.

Cat.	RA	Dec.	Mp	Type	Comments
NGC.4621 M59	$12^h40.8$	$11^o47'$	11.5	E5	

(16) Bright oval much brighter to the middle. About 2' × 1' oriented approximately N/S. Star mag. about 12 1' N.

(8) Small oval patch with PA of about 10°. Brighter nucleus is apparent but not marked. Quite bright and easily seen.

(6) Oval 3' × 2', diffuse halo suddenly brighter to a core.

(20x50) Bright, large, oval.

Cat.	RA	Dec.	Mp	Type	Comments
NGC.4638	$12^h41.6$	$11^o35'$	12.5	E5	See appendix.

(16) Bright with high surface brightness and much brighter to the middle. Small and slightly elongated (about 1' × 30" with PA about 70°. NGC.4637 preceding not found although RNGC magnitude given as 12.0.

(10) Difficult, elongated SSE to NNW. NGC.4637 very close to NW edge of 4638.

(8) Very small and faint. Elongated.

(5) [Misidentified to 4637.] Bright elliptical with a star south following.

	Cat.	RA	Dec.	Mp	Type	Comments
	NGC.4647	$12^h42.3'$	$11^o43'$	12.0	Sc	Double in.
M60 =	NGC.4649	$12^h42.4'$	$11^o42'$	10.5	E2	Apparent contact.

(16) Very bright, large. Bright central region with intense stellar core. Overall about 3' × 2.5'. Slightly elongated with PA approximately E/W. Diffuse halo. Galaxy visible in 60 mm finder quite easily. NGC.4647 NP by about 1½' with a gap of about 1'. Bright and quite large (about 1' diameter) irregularly round and brighter to the middle. Diffuse.

(10) M60 bright, round and much brighter to the middle. 4647 very close NF, a faint blur.

(8) Almost perfectly round very bright core with an almost stellar nucleus. 4647 close NP is faint and irregularly round being brighter to the middle.

(6) M60 is 5' in diameter and slightly elongated with a bright stellar nucleus. NGC.4647 is attached to the N, being a faint featureless oval.

(5) M60 is a very bright, large oval, brighter to the middle. NGC.4647 is closer to the north appearing as a bright, small oval with a soft texture.

Cat.	RA	Dec.	Mp	Type	Comments
NGC.4654	$12^h42.7'$	$13^o16'$	11.5	S	IC3708.

(5) Bright oval, very opaque.

Cat.	RA	Dec.	Mp	Type	Comments
NGC.4689	$12^h46.6'$	$13^o54'$	12.0	S	

(5) Faint oval, opaque.

Cat.	RA	Dec.	Mp	Type	Comments
NGC.4698	$12^h47.2$	$08^o38'$	12.0		Spindle.

(5) Faint, small, almost stellar.

Cat.	RA	Dec.	Mp	Type	Comments
NGC.4754	$12^h51.1'$	$11^o27'$	12.0		

(16) Bright and much brighter to the middle. Round, about 1' in diameter. NGC.4762 following.

(8) Round, bright and brighter to the middle. Small almost stellar nucleus.

(5) Bright ellipse, soft texture.

Cat.	RA	Dec.	Mp	Type	Comments
NGC.4762	$12^h51.7'$	$11^o22'$	11.5	SO	Edge on.

(16) Very bright and much brighter to the middle to a stellar nucleus. Spindle very much extended in PA of about -30°. About 3' × 30".

(8) A very thin nebula, bright and oriented with PA of 40°. Bright nucleus also elongated. Overall 4' × $\frac{1}{2}$'.

(5) Very long bright streak with a wide middle.

Appendix: The Confusion over NGC.4657 and NGC.4638

The RNGC places two bright galaxies at the positions $12^h41.7$, $11^o38'$ $12^h41.6$, $11^o35'$ (1975): they are respectively RNGC.4637 and 4638. These are described by the authors from the POSS as

'4637 - Elongated, faint, diffuse, brighter to the middle, 38 near.
4638 - Edge on, brighter to the middle, 37 near.'

In the original NGC published in the memoirs of the Royal Astronomical Society, the two objects are described as follows. NGC.4638, discovered by William Herschel and noted twice as II70 and II176, has the NGC description - faint, round and greatly brighter to the middle. NGC,4637, which first appeared in the General Catalogue of John Herschel is described as "makes a double nebula with h1402' (NGC.4638) and is placed SP 4638". In the notes section, Dreyer adds the following comments. "It is very possible that the Birr observer mistook M60 and III44 for h1402 and a nova. Schultz says h1402 is conspicuously binuclear. Why was this not noticed by John Herschel, d'Arrest or by Vogel?".

It is unfortunate that the RNGC does nothing to clear up this confusion, being based upon inspection of the POSS. The magnitudes and descriptions give the impression of a conspicuous double galaxy. However, this is manifestly not the case. Examination of the 12" photograph on p. 78 of Newton's 'Deep Sky Objects"[3] reveals only the object sketched using a 16 inch shown in this volume. The photograph which appears in Kreimer's 'Messier Album'[4] also shows a single galaxy. To settle this matter, reference was made to the POSS. Just following the bright galaxy (NGC.4638) is a tiny spindle shaped object, only just non stellar, about 20" × 10", possibly an edge-on lenticular, (SO) galaxy. If its magnitude is 12.0, it is strange that the 16 inch observer easily picked up the 14.5 mag. low surface brightness galaxy, 4607, a few minutes afterwards, but missed such a supposedly high contrast 12 mag. almost stellar galaxy. The RNGC magnitude must be in error to account for this. However, J. Perkins reports on observation with a 10 inch in this volume. Its visibility must therefore remain the object of contention.

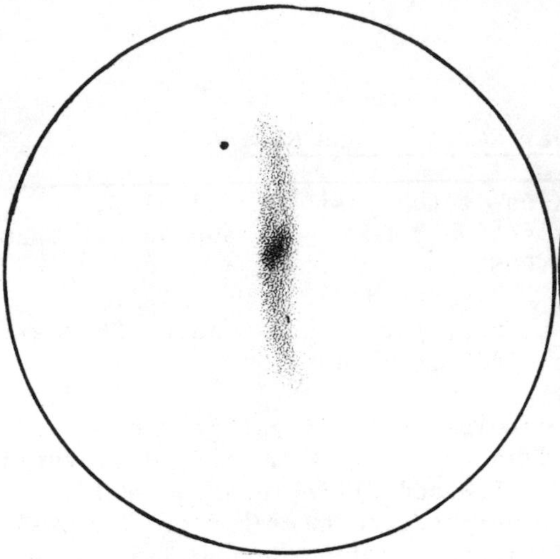

NGC 4192/M98
GSW 16-inch (12')

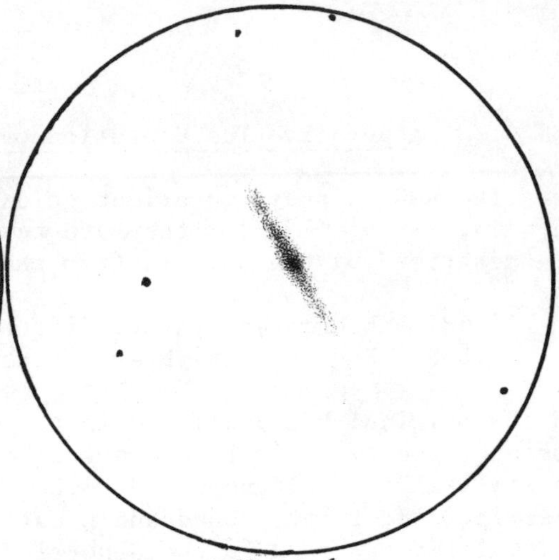

NGC 4216
GSW 16-inch (12')

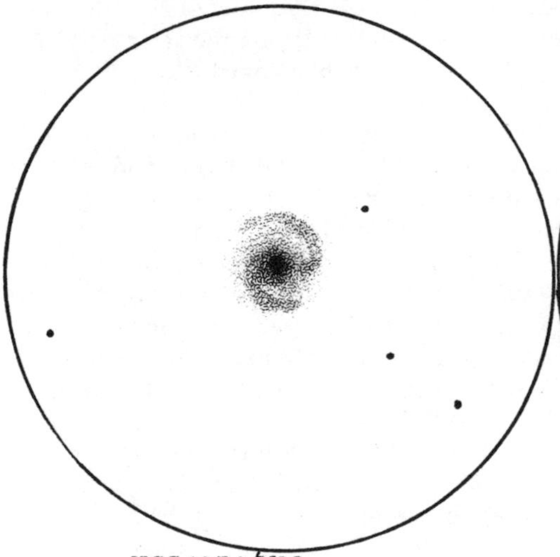

NGC4254/M99
GSW 16-inch (12')

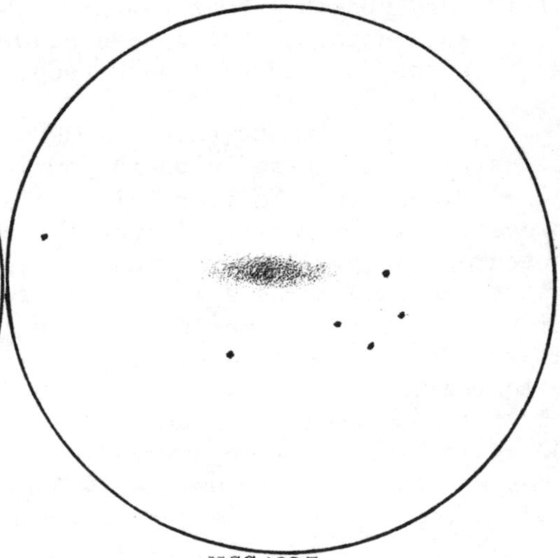

NGC4293
GSW 16-inch (12')

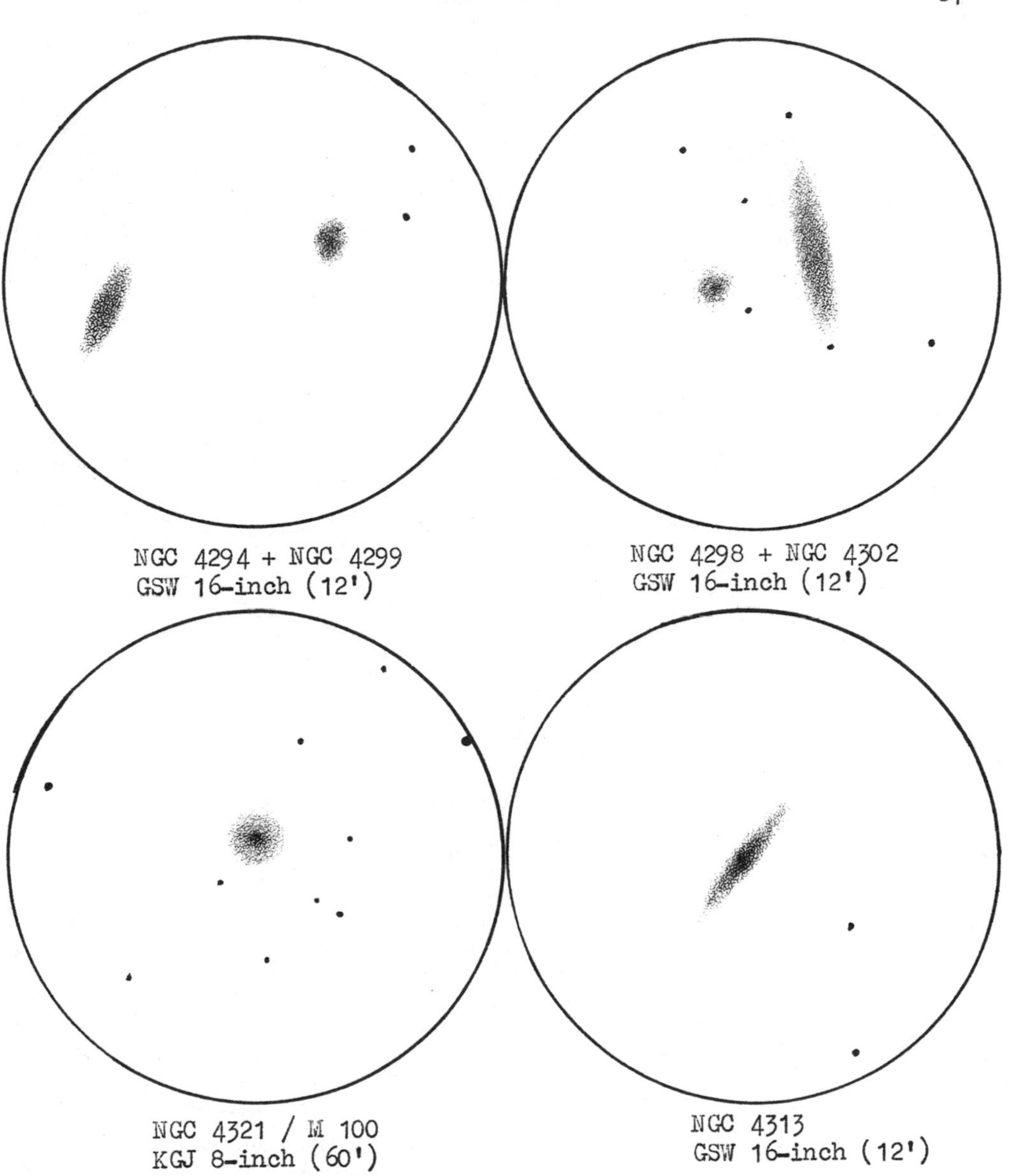

NGC 4294 + NGC 4299
GSW 16-inch (12')

NGC 4298 + NGC 4302
GSW 16-inch (12')

NGC 4321 / M 100
KGJ 8-inch (60')

NGC 4313
GSW 16-inch (12')

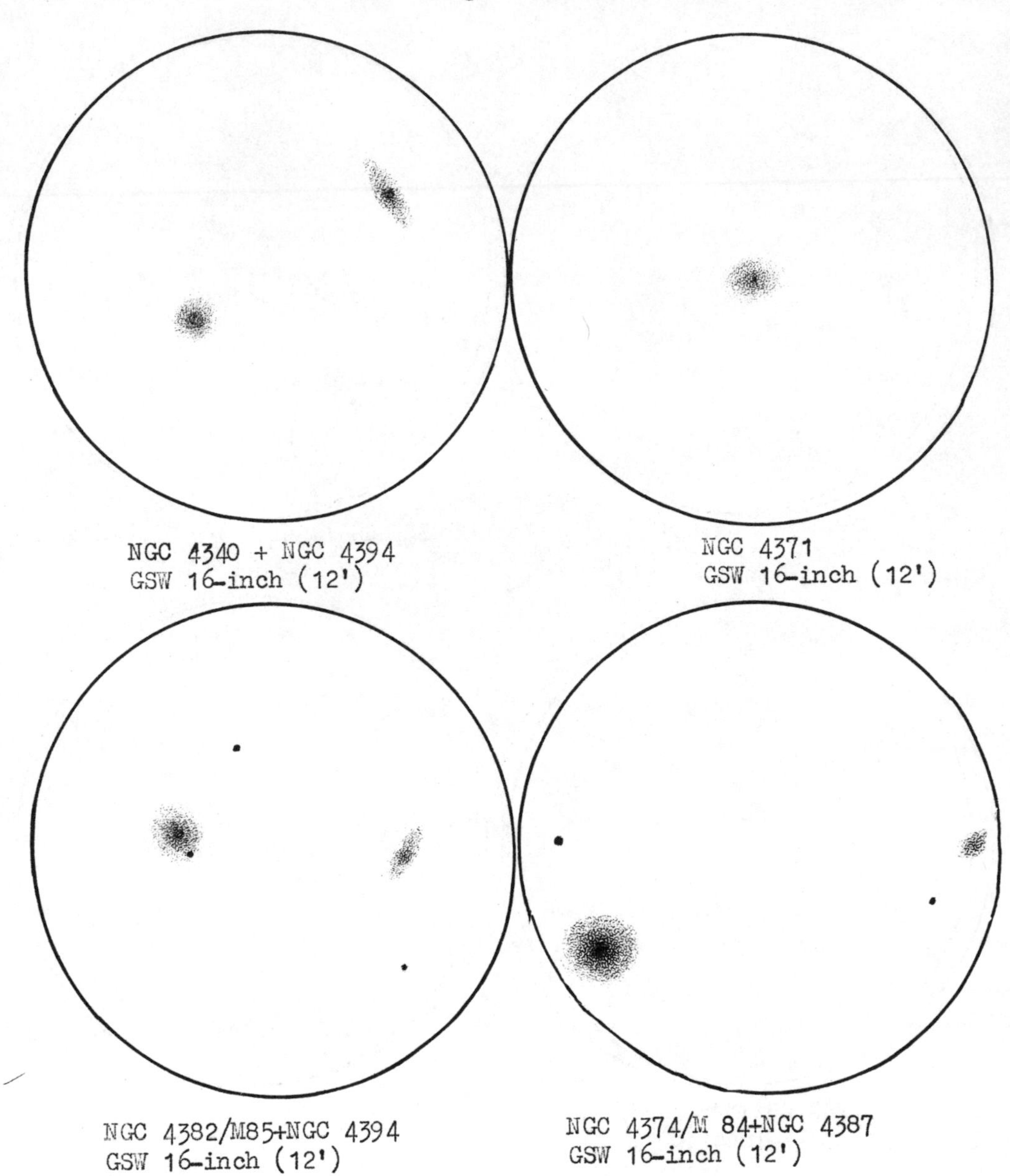

NGC 4340 + NGC 4394
GSW 16-inch (12')

NGC 4371
GSW 16-inch (12')

NGC 4382/M85+NGC 4394
GSW 16-inch (12')

NGC 4374/M 84+NGC 4387
GSW 16-inch (12')

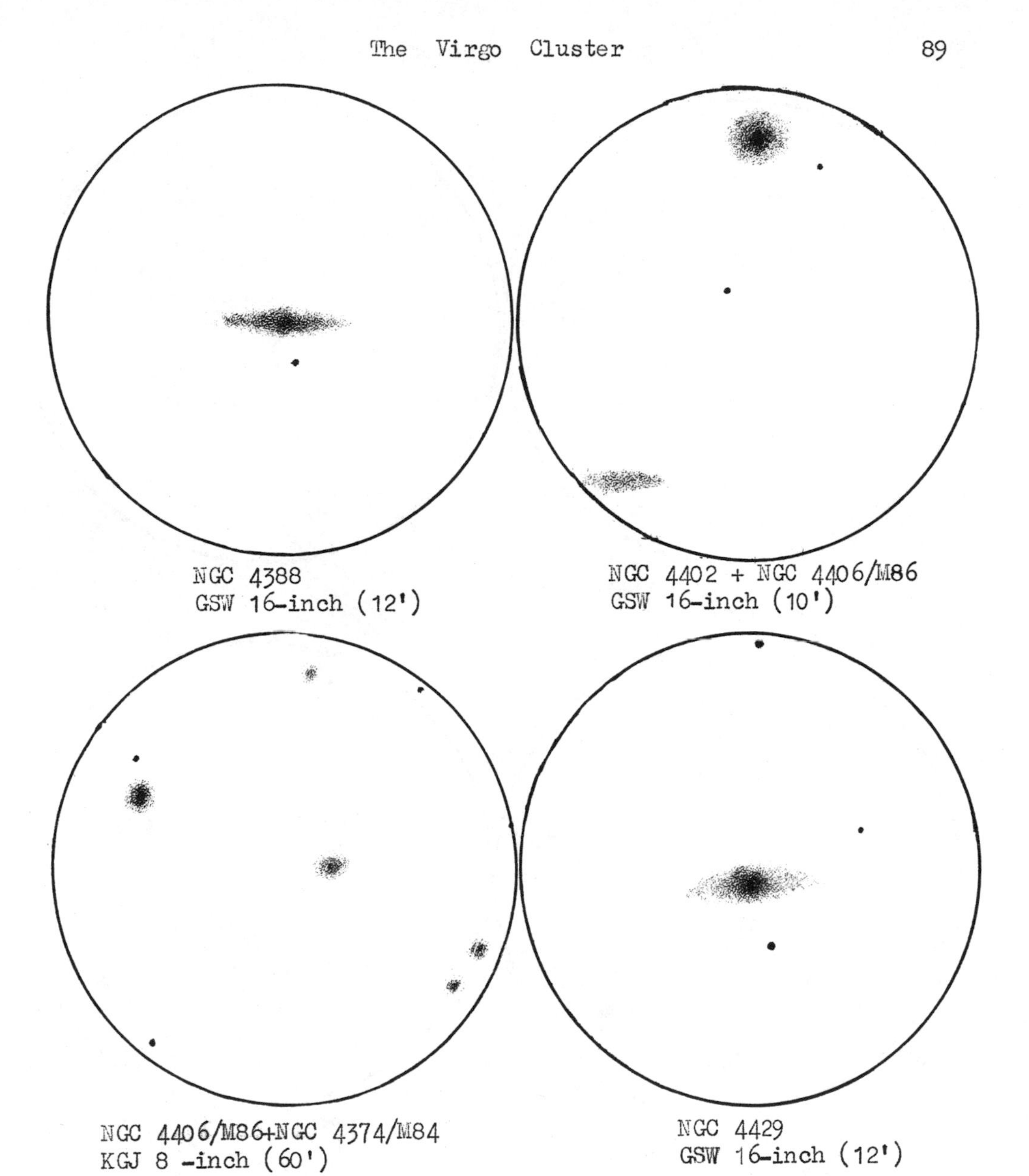

NGC 4388
GSW 16-inch (12')

NGC 4402 + NGC 4406/M86
GSW 16-inch (10')

NGC 4406/M86+NGC 4374/M84
KGJ 8 -inch (60')

NGC 4429
GSW 16-inch (12')

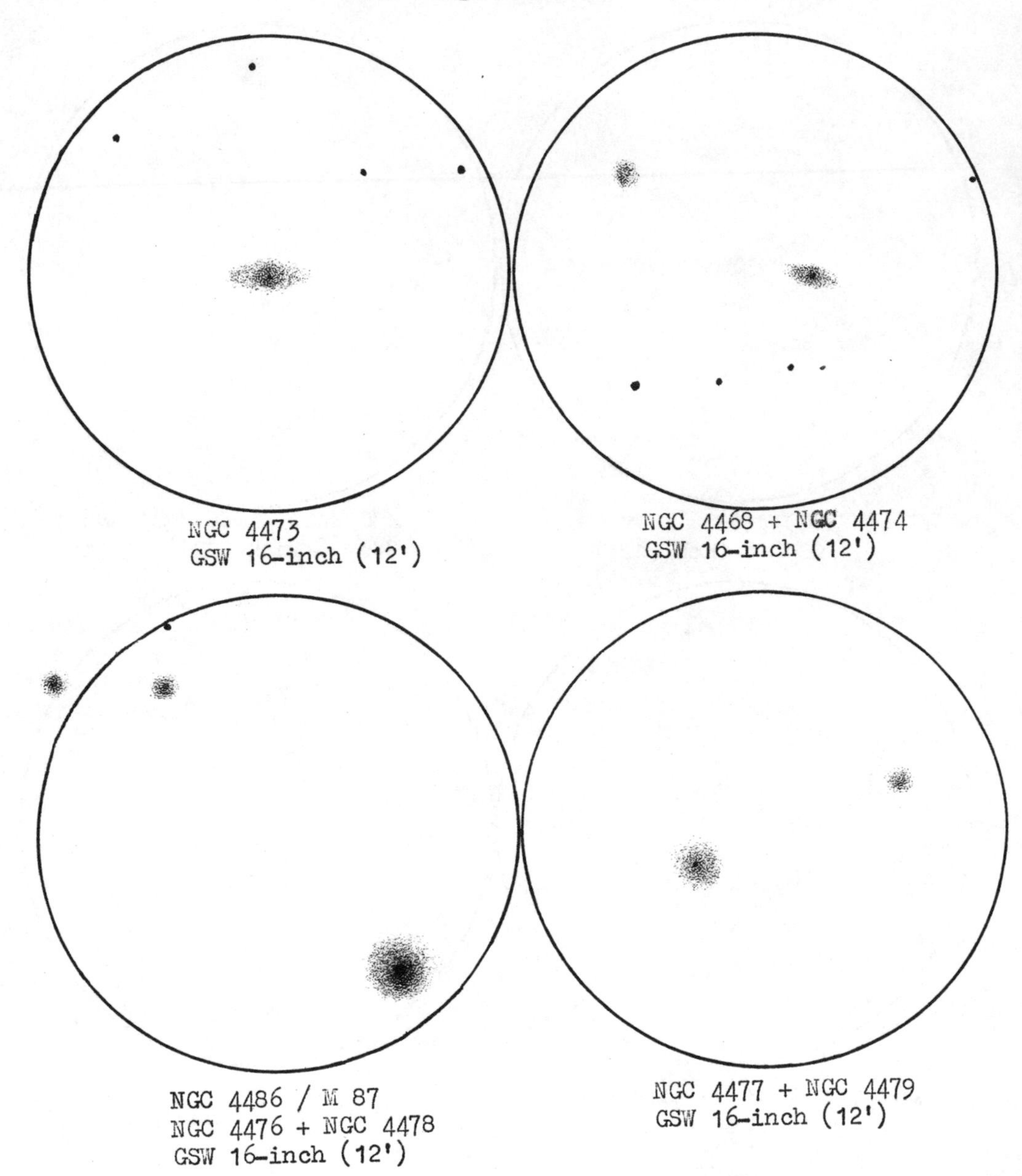

NGC 4473
GSW 16-inch (12')

NGC 4468 + NGC 4474
GSW 16-inch (12')

NGC 4486 / M 87
NGC 4476 + NGC 4478
GSW 16-inch (12')

NGC 4477 + NGC 4479
GSW 16-inch (12')

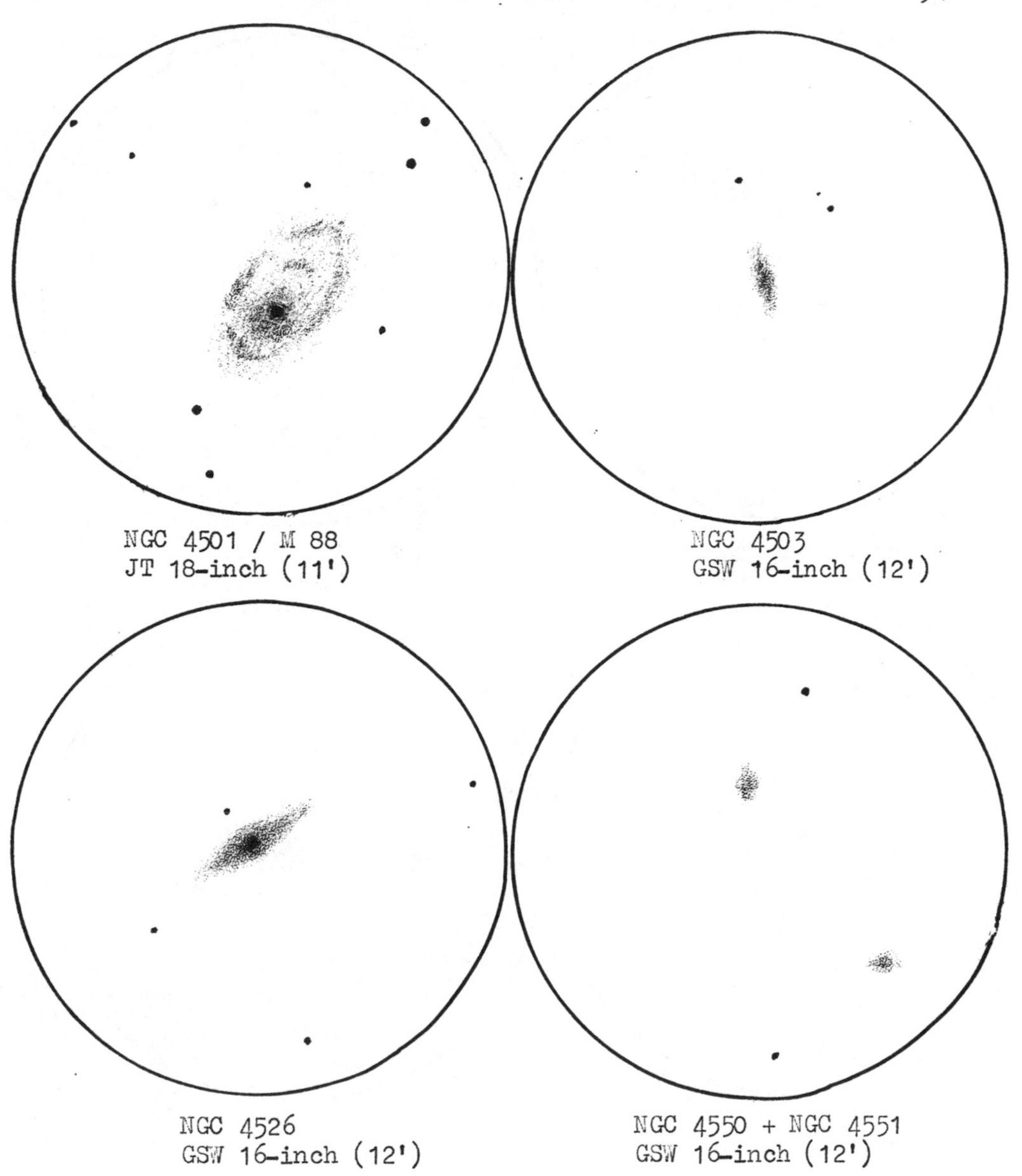

NGC 4501 / M 88
JT 18-inch (11')

NGC 4503
GSW 16-inch (12')

NGC 4526
GSW 16-inch (12')

NGC 4550 + NGC 4551
GSW 16-inch (12')

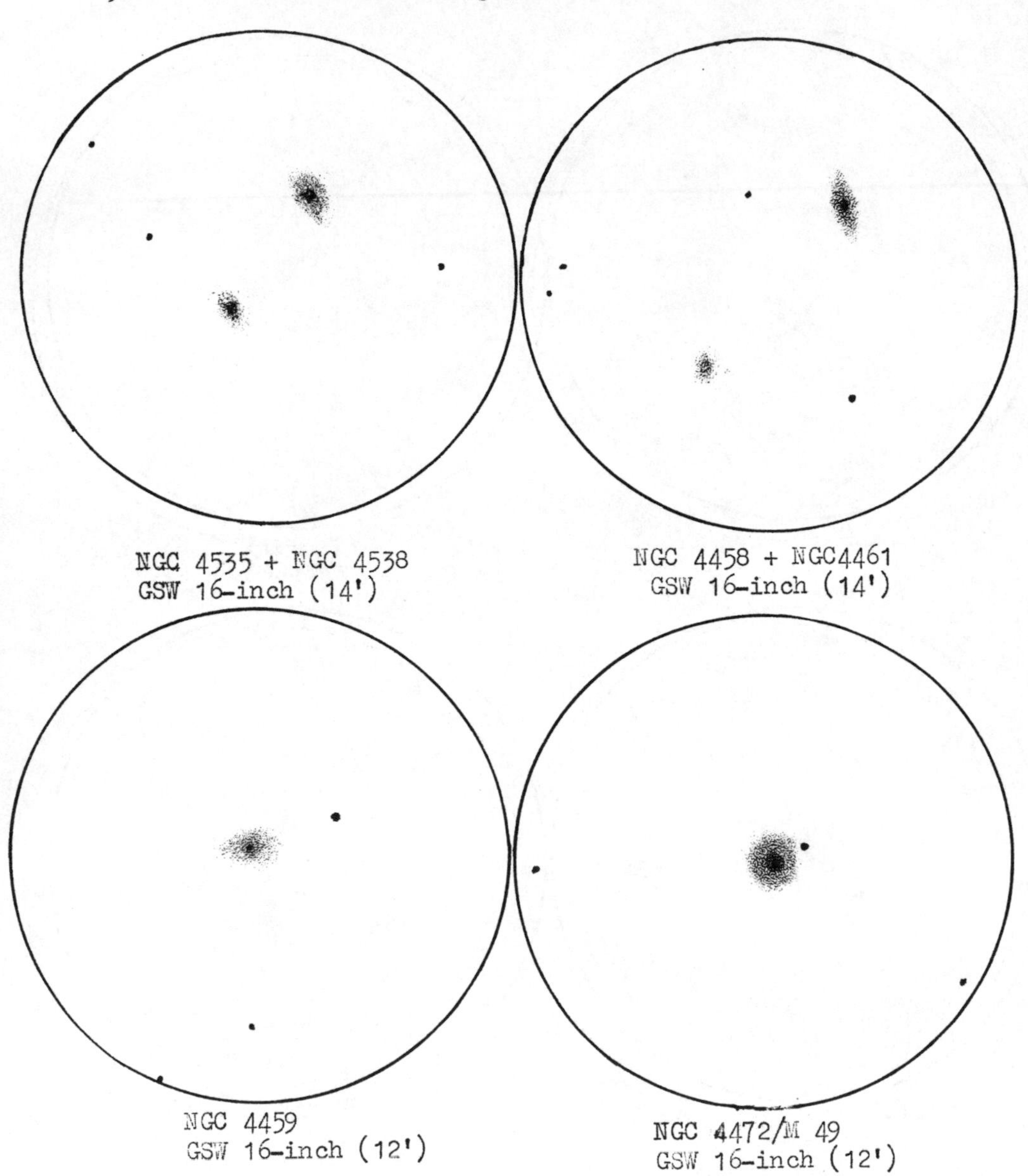

NGC 4535 + NGC 4538
GSW 16-inch (14')

NGC 4458 + NGC4461
GSW 16-inch (14')

NGC 4459
GSW 16-inch (12')

NGC 4472/M 49
GSW 16-inch (12')

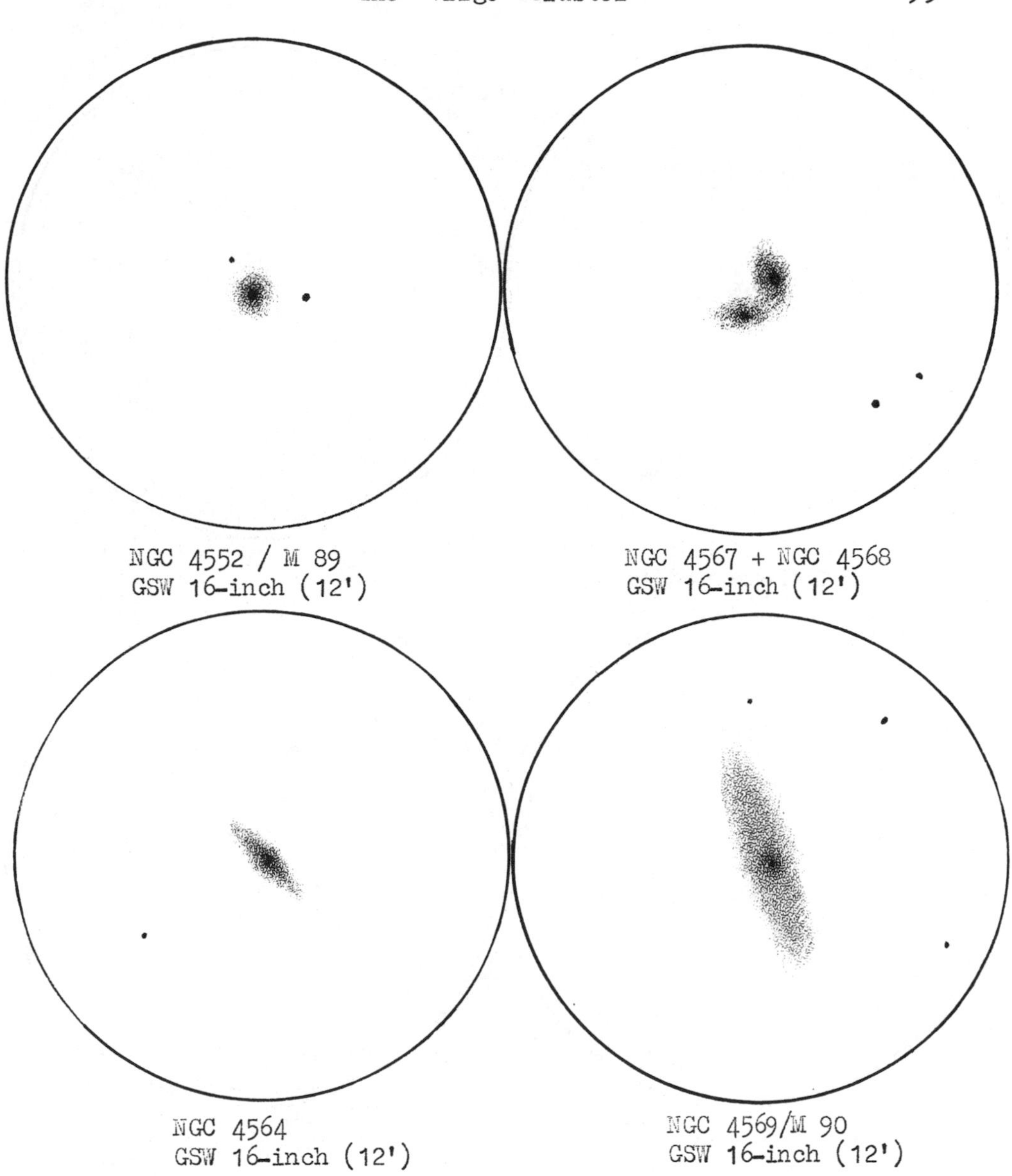

NGC 4552 / M 89
GSW 16-inch (12')

NGC 4567 + NGC 4568
GSW 16-inch (12')

NGC 4564
GSW 16-inch (12')

NGC 4569/M 90
GSW 16-inch (12')

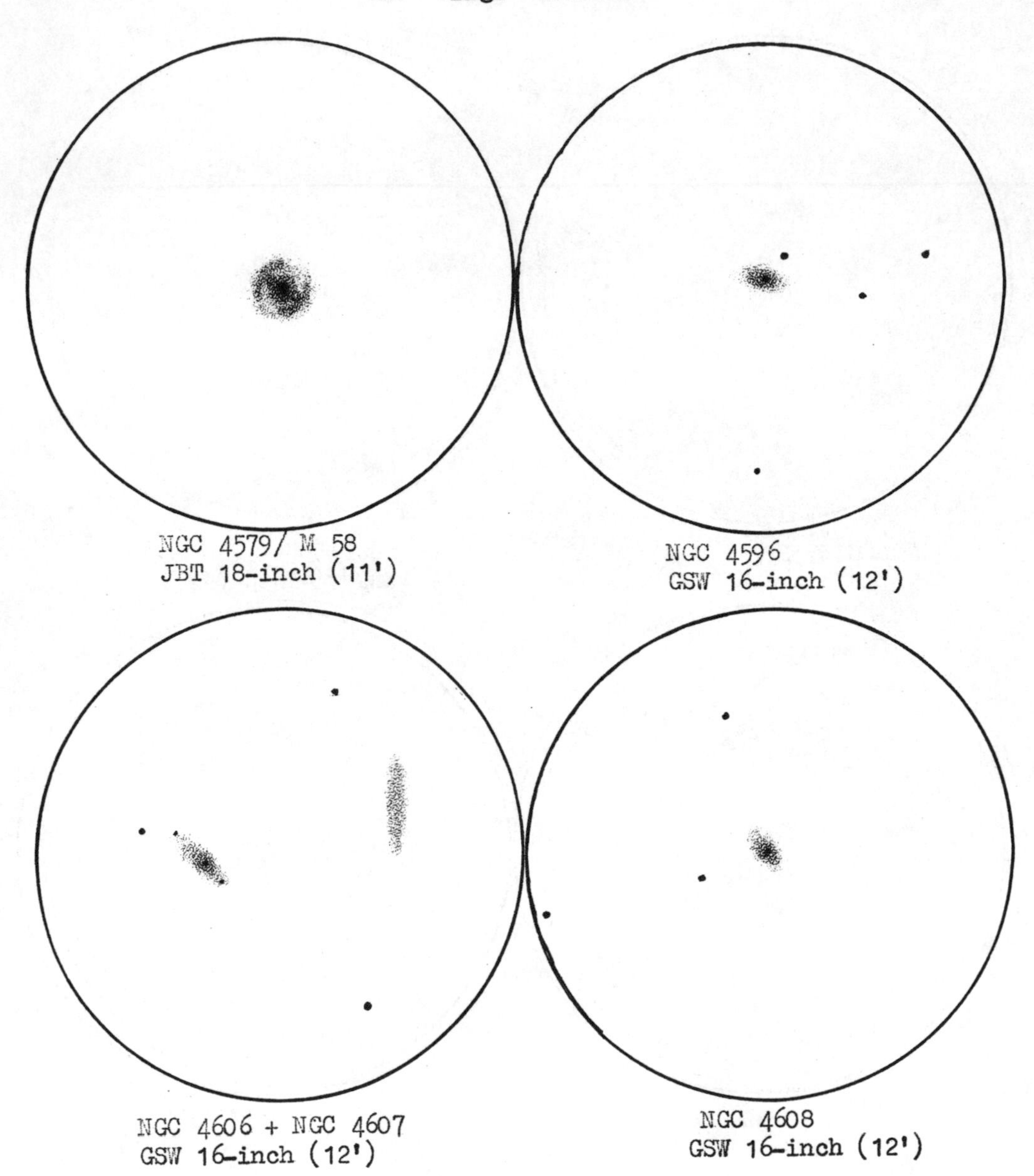

NGC 4579/ M 58
JBT 18-inch (11')

NGC 4596
GSW 16-inch (12')

NGC 4606 + NGC 4607
GSW 16-inch (12')

NGC 4608
GSW 16-inch (12')

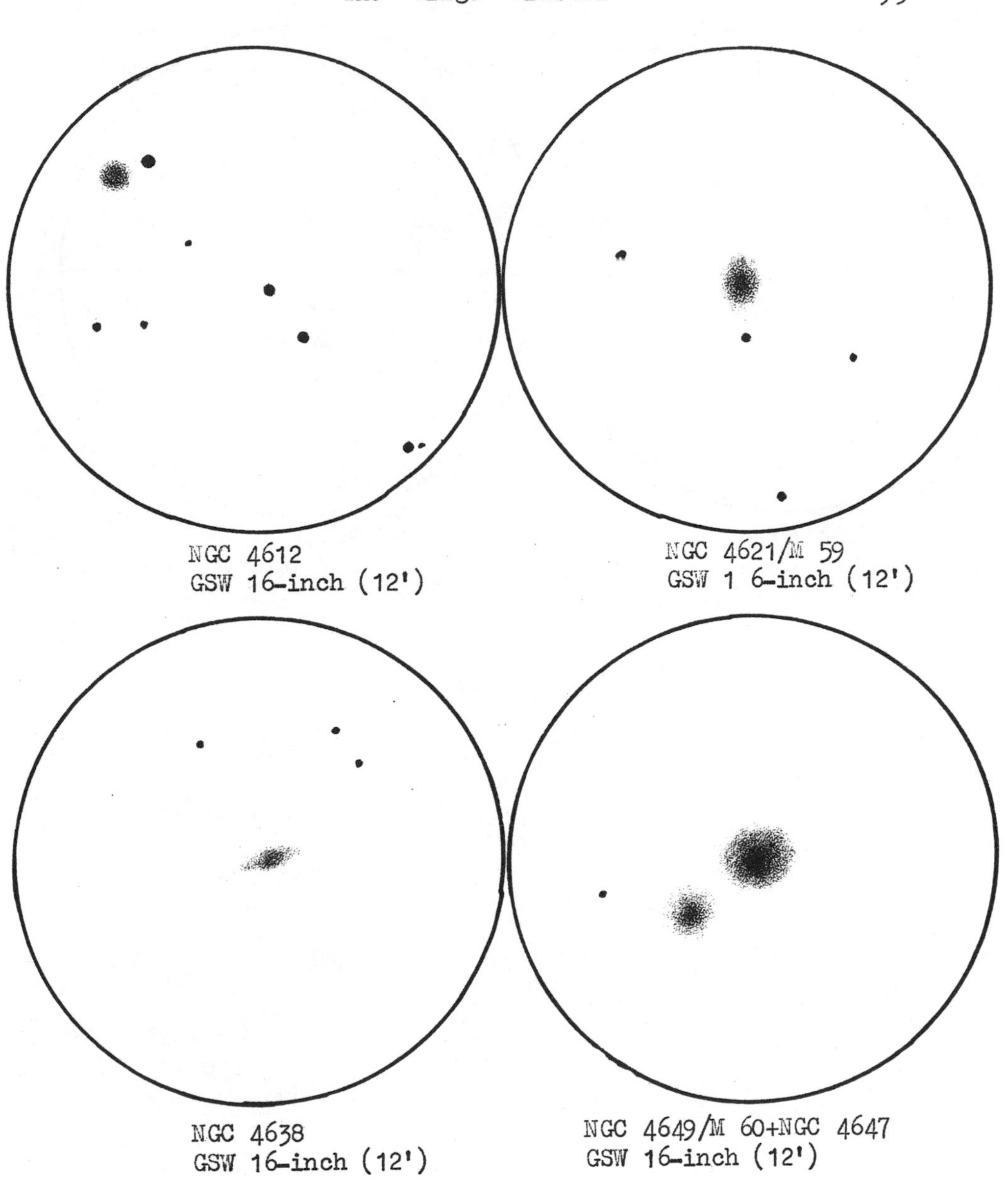

NGC 4612
GSW 16-inch (12')

NGC 4621/M 59
GSW 1 6-inch (12')

NGC 4638
GSW 16-inch (12')

NGC 4649/M 60+NGC 4647
GSW 16-inch (12')

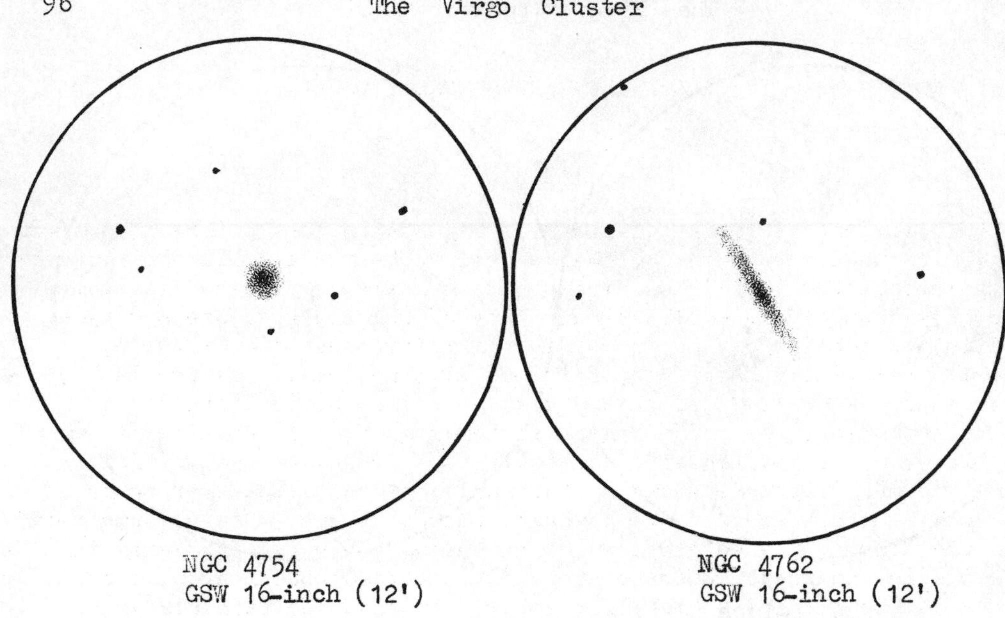

NGC 4754
GSW 16-inch (12')

NGC 4762
GSW 16-inch (12')

2. ABELL CLUSTERS OF GALAXIES

INTRODUCTION

The following catalogue contains visual observations of Abell clusters of galaxies made by a number of Webb Society observers. All the clusters were selected from Abell's catalogue on the basis of distance (distance group less than 4) and declination. Many of the clusters are not rich visually, this being a function of intrinsic richness and distance. Indeed, a relatively large proportion of all the Abell clusters in distance groups 0, 1, 2 and 3 have richness class 0, indicating a population of less than 50 galaxies in the magnitude interval m_3 to $m_3 + 2$, whilst some of the richer clusters lie at true distances sufficient to dim most of their galaxies out of the range of visual observation in almost all amateur apertures. However, the catalogue contains some very rewarding clusters, even amongst the richness 0 objects. For example, Abell 194 is one where moderate amateur apertures should reveal over a dozen galaxies with rich low power galaxy fields. The catalogue also contains 5 compact richness class 2 clusters A426 (Perseus), A1656 (Coma), A1367, A2151 and A2199. The first three of these probably contain the richest visual galaxy fields to be found in the whole sky, with A1656 being the most beautiful of all. The clusters A426 and A1367 contain fields revealing up to nine galaxies in the 15' MP field of a 16 inch Newtonian, but Coma contains several such rich fields whilst the field of NGC.4874 contains 18 galaxies.

Apart from the pleasure to be derived from the observation of these crowded galaxy fields containing untold billions of stars, the rich clusters of galaxies can offer the satisfaction of the 'visual discovery' of anonymous galaxies - a pastime often extolled by Malcolm Thomson. An 'anonymous galaxy' is defined as one which does not appear in the NGC or IC. Often these galaxies are listed in the more complete catalogues such as the MCG, UCG or CGCG. However, even these catalogues are incomplete for magnitudes greater than 15. Owners of larger amateur apertures can therefore expect to pick up 'completely anonymous' galaxies not appearing in any of the standard catalogues. Although all the galaxies will appear on the POSS or in more detailed plates, the visual observer of these faint objects can be fairly certain that he or she will be amongst the very few, or even be the first, to have observed the objects visually.

Another source of inspiration associated with the visual observation of distant clusters of galaxies is the realisation that one is contemplating the large end of the scale of things. Clusters of galaxies themselves probably represent the largest physical subsystems of matter in the universe. They also lie at colossal distances. It is interesting to consider the scale of distance covered by the field of view of one's eyepiece as one looks into the depths of cosmic space using the galaxies in clusters as reference points. Suppose, for example, that one is using an eyepiece with a field of 15' of arc. Then, at the distance of the Virgo cluster (20 Mpc) one's field of view has a true diameter of about 0.1 Mpc, whilst at the distance of the Ursa Major I Cluster (A1377) (306 Mpc) the field has a true diameter of about 1.3 Mpc, about twice the distance to the Andromeda Nebula!

With the above motivations in mind, we now turn to a description of the catalogue. Each cluster is treated in a similar way to the Virgo cluster. First of all, there is a summary of classificational and positional information. This is followed by a brief description of the cluster, together with two finder charts. The first is a large scale chart for location of the cluster field amongst the stars, and the second is a chart, usually about $1^o \times 1^o$, of the central region of the cluster with galaxy identifications. These are drawn from various sources. Preference is given to identifications quoted in research papers, whilst if these are not available, the brighter galaxies are identified from the Morphological Catalogue of Galaxies. In a few places, the NGC or IC identifications are ambiguous - we have already noted such an ambiguity in the case of NGC.4637/4638 in the Virgo Cluster. The POSS-based RNGC does not always help here - in some cases it adds to the confusion. Wherever possible, such ambiguities are pointed out and solutions suggested by cross-reference to catalogues and plate material.

The charts of the central regions were prepared by various methods, and as these may introduce confusion between almost stellar galaxies and faint stars, reliance should be placed only upon the galaxies of appreciable angular size. As many field stars as possible are included, as these are useful reference points for visual work at the telescope and for later identification purposes. The large-field finder charts of regions around cluster centres were copied by hand from the SAO atlas, and the NGC and IC field galaxies were marked in, using MCG positions. Although it would possibly have been useful to mark in all galaxies brighter than magnitude 15, in many cases this would have led to overcrowding, and this was omitted in all cases (except A194).

Each cluster description will be preceded by a table of information containing (1) The Abell Catalogue Number (2) RA and Dec (1950.00) (3) Abell Distance Class (4) Abell Richness Class (5) mean redshift and (6) Rood Sastry Class. Following the descriptive outline is a list of cluster galaxies in the central region as depicted upon the chart. The list contains positional data, photographic or visual magnitudes and cross-identifications between the NGC, IC and MCG or CGCG where appropriate. In the case of some of the more distant clusters, photometric and identification data from research papers are provided, as many of the galaxies do not appear in any large catalogue.

LIST OF CLUSTERS WITH OBSERVERS

The catalogue contains observations of the following rich clusters of galaxies made by the observers indicated.

Abell	RA	Dec.	Observers
119	$00^h53.9'$	$-01^o32'$	GSW
194	$01^h32.0'$	$-01^o38'$	GSW
262	$01^h49.9'$	$+35^o55'$	RB,MJT,GSW
347	$02^h22.0'$	$41^o44'$	MJT,GSW
426	$03^h12.0'$	$41^o22'$	DB,ESB,RB,MJT,GSW
1185	$11^h08.2'$	$28^o57'$	GSW
1228	$11^h18.9'$	$34^o37'$	GSW
1367	$11^h41.9'$	$20^o07'$	RB,MJT,GSW
1377	$11^h44.4'$	$56^o01'$	GSW
1656	$12^h57.4'$	$28^o15'$	ESB,RB,JP,MJT,GSW
2065	$15^h20.0'$	$27^o50'$	RB
2147	$16^h00.0'$	$16^o03'$	GSW
2151	$16^h03.0'$	$18^o00'$	RB,GSW
2197	$16^h26.5'$	$41^o01'$	GSW
2199	$16^h26.9$	$39^o38'$	RB,MJT,GSW

List of Observers Instruments and Locations	
E.S. Barker	8½ inch, Herne Bay, UK
D. Branchett	5 inch, Eastleigh UK
R. Buta	107, 82, 36, 30 inch McDonald Observatory, Texas, USA
J. Perkins	10 inch, Kirby-in-Ashfield, UK
M.J. Thomson	16½ inch, Santa Barbara, USA
G.S. Whiston	16 inch, Witley, UK

CATALOGUE OF VISUAL OBSERVATIONS

1. Abell 119

Abell			AD	AR	Z	RS
119	$00^h53.9'$	$-01^o32'$	3	1	0.045	C

Abell 119 is one of the most distant clusters in this catalogue although its visual observation in moderate amateur apertures should not present much difficulty, its brightest galaxies having apparent photographic magnitude around 15. These are MCG 0-3-34 and MCG 0-3-33 marked as L and H on the chart of the central region. On a deep 4 m plate of the cluster(1) the latter galaxy shows an extended halo in and around which are many dwarf galaxies. Abell 119 has a velocity dispersion of around 1000 km sec^{-1} and has been detected at X-ray and radio wavelengths. Evidence has also been produced to show that the percentage of E and SO galaxies to spirals increases towards the cluster core. As we have seen, these are all characteristics associated with a dense intracluster medium.

List of Catalogued Galaxies in Abell 119

The positions and photographic magnitudes quoted in the following table were taken from the CGCG and the cross-correlated with the MCG and the UCG.

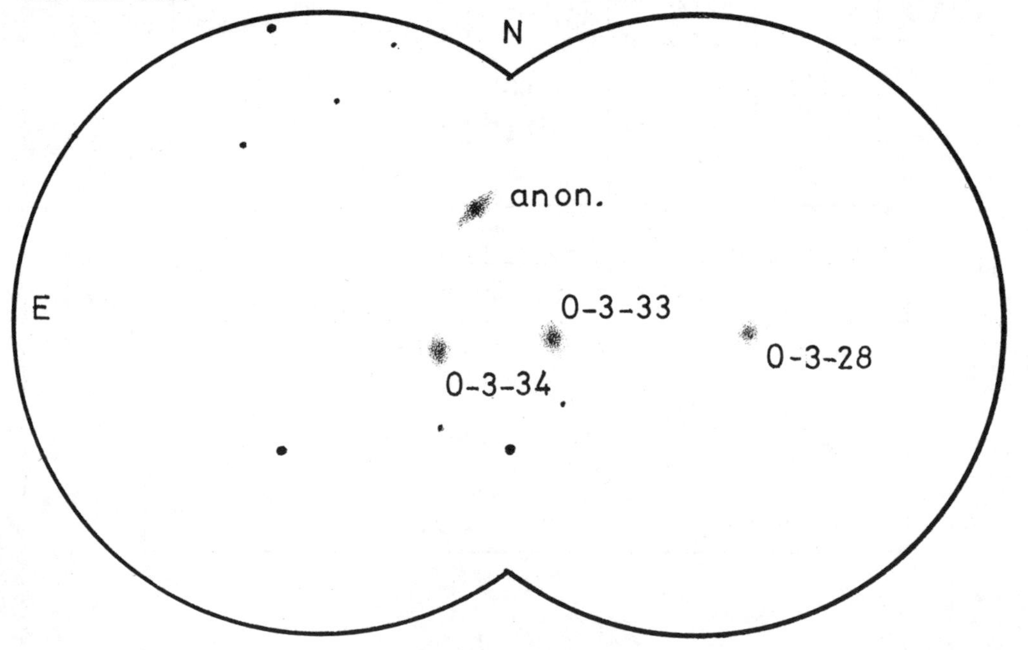

ABELL 119 GSW 16-inch x160 field 15'

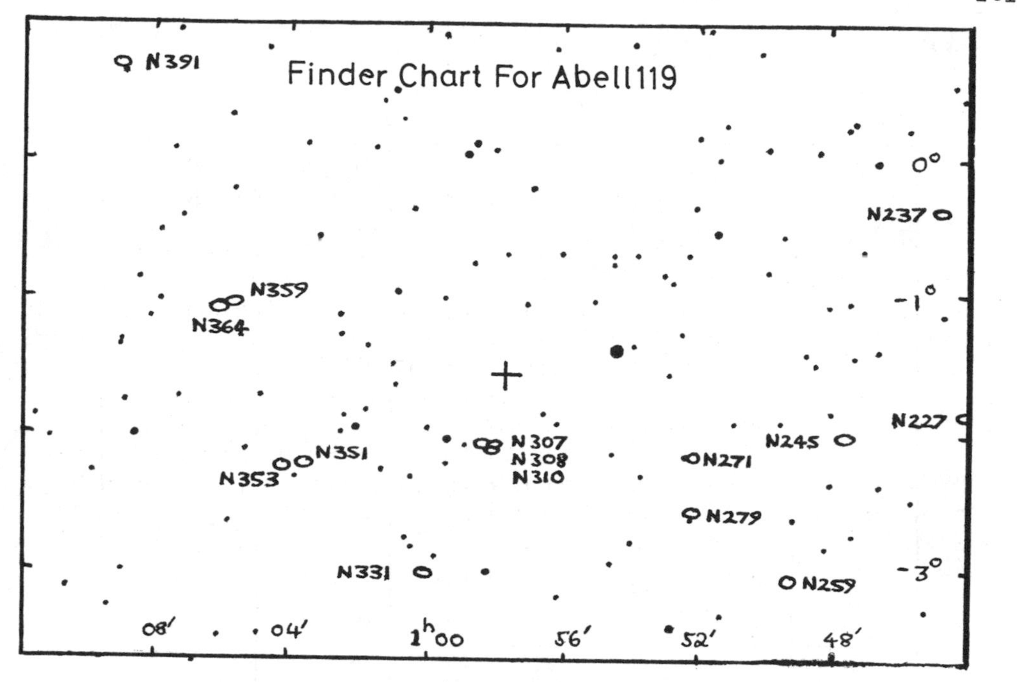
Finder Chart For Abell119
N391
N359
N364
N237
N227
N245
N307
N308
N310
N351
N353
N271
N279
N331
N259
0°
-1°
-3°
08′
04′
1h00
56′
52′
48′

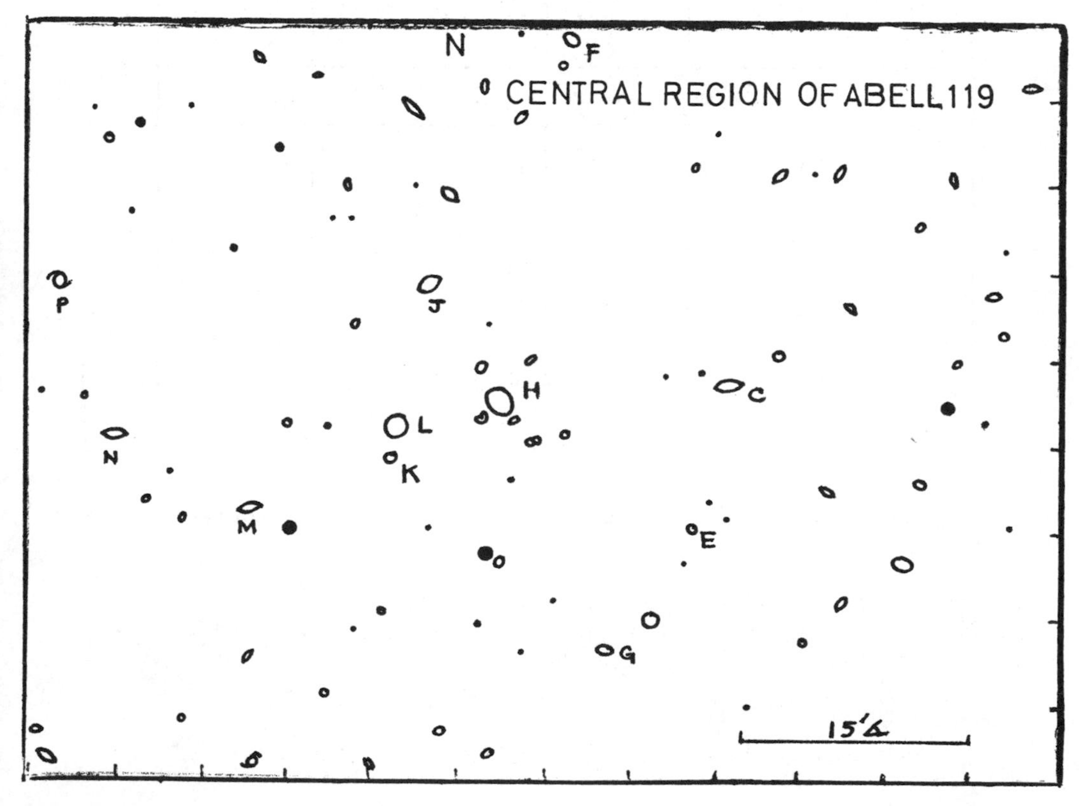
CENTRAL REGION OF ABELL119
N
F
P
J
H
C
L
K
N
M
E
G
15′.4

Chart	RA	Dec.	Mp	MCG	UCG or other
C	$00^h52.8'$	$-01^o31'$	15.2	0-3-28	
E	53.2'	33'	15.7		
F	53.4'	10'	15.3	0-3-29	00570
G	53.5'	35'	15.4	0-3-32	
H	53.7'	30'	15.0	0-3-33	00579
J	53.8'	28'	15.3		
K	53.9'	32'	15.6		
L	53.9'	31'	14.9	0-3-34	00583
M	54.1'	33'	15.5	0-3-37	
N	54.3'	31'	15.4	0-3-38	
P	54.4'	28'	15.3		
The following objects are not in the above field					
	$00^h51.7'$	$-01^o41'$	15.7		00568
	52.7'	18'	15.7	0-3-25	
	54.5'	07'	15.1	0-3-39	
	55.1'	38'	14.8		3C29
	55.4'	03'	15.7		

Visual Observations

The cluster was easily picked up using a 16 inch Newtonian and four galaxies observed in the central region.

MCG 0-3-28 First of 4 galaxies at Mp field. A small faint object preceding a compact group of 3 galaxies. Irregularly round (indefinite PA).

MCG 0-3-33 Brightest in field and first of a compact group of 3. Quite bright overall and suddenly much brighter to the middle. Slightly elongated, perhaps N/S and about 25" × 15" in size.

(J) Quite bright, slightly elongated N/S.

MCG 0-3-34 Second brightest in field. Similar to MCG 0-3-33 in size and overall appearance but slowly brighter to the middle. Slightly elongated parallel to MCG 0-3-33.

2. Abell 194

Abell	RA	Dec.	AD	AR	Z	RS
194	$01^h23.0'$	$-01^o38'$	1	0	0.018	L

Abell 194[1] has Abell richness 0, but because it is so close (redshift distance 110 Mpc for H_o = 50 km sec^{-1} Mpc^{-1}) it is a rewarding object for visual observation. As the Rood-Sastry class implies, the cluster is highly elongated, with most of the brighter galaxies lying roughly on a line joining the apparently overlapping pair of galaxies NGC.545/547 to NGC.541. The former pair of galaxies is the brightest visual nebula in the cluster, whilst NGC.541 lies at the centre of the cluster as defined by Zwicky and Humason. Deep photometry[9] shows that a faint luminous bridge, probably composed of stars, joins NGC.545/547 to NGC.541 and embedded in the bridge is a peculiar galaxy known as Minkowski's Object. The latter is marked 'M' on the chart. Its optical spectrum is rather strange, containing some emission lines found in extragalactic HII regions usually associated with peculiar elliptical galaxies, whilst other lines are reminiscent of lines usually associated with N. galaxies. Because of the near presence of Minkowski's Object, NGC.541 appears in Arp's atlas of peculiar galaxies as Arp133. NGC.545/547 also appears as Arp308 being identified as a radio source (3C40) which has a very complex multiple structure, with an intensity peak centred on Minkowski's object. A 48-inch Schmidt Plate of the Central Region is reproduced in[3].

List of Catalogued Galaxies in Abell 194

The following data is taken from[1] and contains positional data due to Zwicky and Humason, photographic magnitude and estimated Hubble types. The Zwicky Humason identifications are cross-correlated with the NGC and MCG in the last columns.

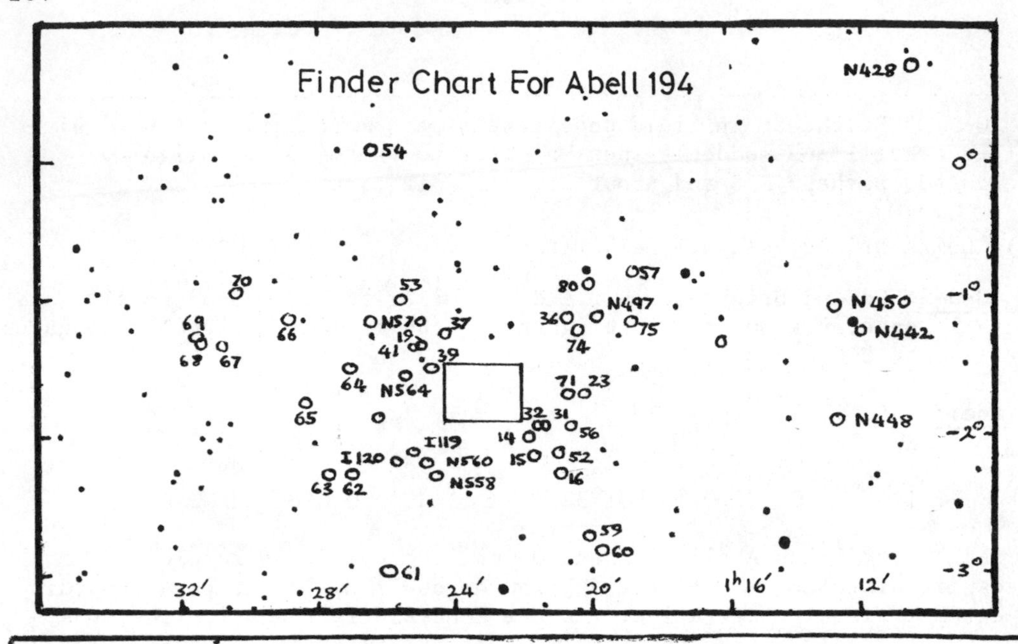
Finder Chart For Abell 194
N428
054
70
53
057
80
N497
N450
69
66
N570
37
36
75
N442
68
67
19
41
39
74
64
N564
71
23
65
32
31
N448
56
I119
14
I120
N560
15
52
63
62
N558
16
59
60
61
32′
28′
24′
20′
1h16′
12′
−1°
−2°
−3°

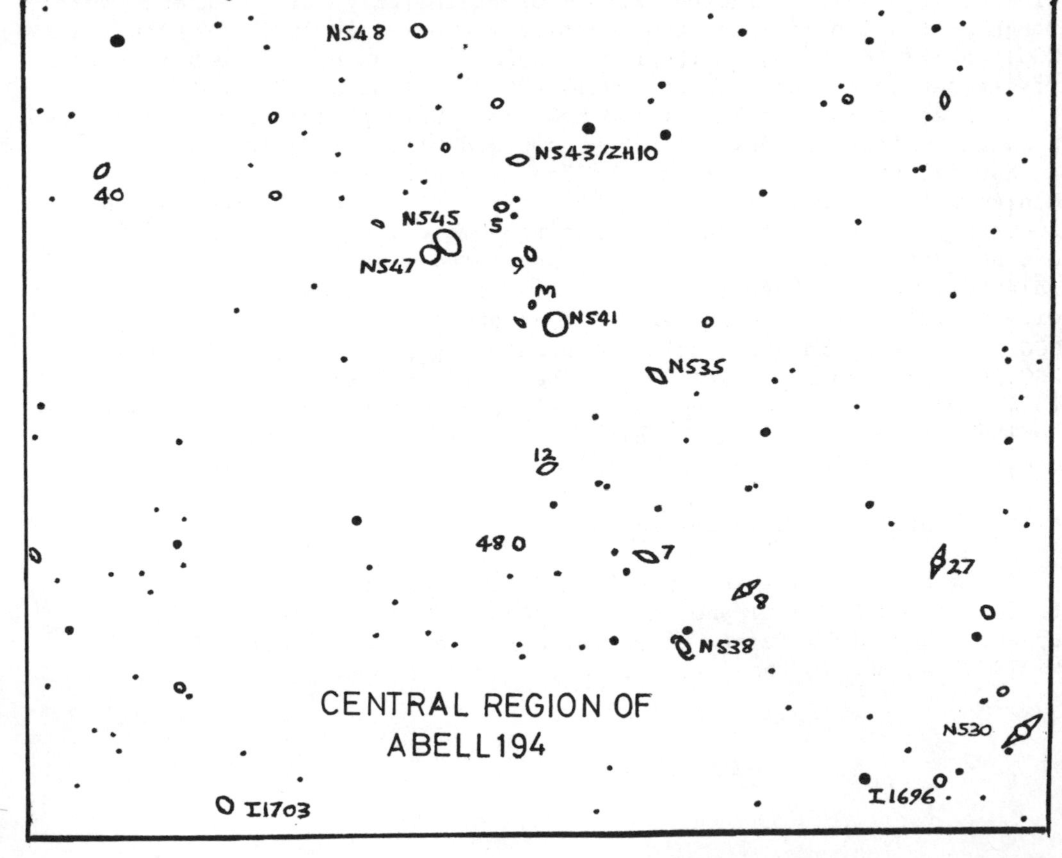
N548
N543/ZH10
40
N545
5
N547
9
M
N541
N535
12
48
7
27
8
N538
CENTRAL REGION OF
ABELL194
N530
I1696
I1703

ZH	RA	Dec.	Mp	Type	NGC/IC	MCG
58	$01^h16.4'$	$-01^o16'$	15.0	SBO		0-4-77
57	18.8	48'	14.6	SBb		0-4-95
75	18.8	10'	15.0	SBcdp	I1681	0-4-97
22	19.8	08'	14.1	Sc	N497	0-4-100
59	20.0	$-02^o40'$	15.2	SO		
80	20.1	$-00^o50'$	15.2	SO?		0-4-101
23	20.2	$-01^o39'$	14.6	So		0-4-104
74	20.3	$-00^o13'$	15.4	SO		
71	20.7	$-00^o39'$	15.0	SBc		0-4-107
56	20.7	$-00^o54'$	15.3	SO		0-4-108
36	20.7	$-01^o09'$	15.5	SO/a?		
16	20.8	$-02^o14'$	14.8	SBab		0-4-112?
52	21.1	$-02^o05'$	15.4	E3?		
31	21.4	-01^o53	15.5	E6		
32	21.5	$-01^o54'$	15.3	SOp		
15	21.6	$-02^o07'$	14.7	Sa?		0-4-114
14	21.8	$-01^o00'$	14.9	SO/a	N519?	0-4-116
28	21.9	$-01^o39'$	15.7	SBO?		
30	21.9	$-01^o53'$	15.3	E1		
13	22.1	$-01^o50'$	14.5	SBO	I106	0-4-119/I106
24	22.3	$-01^o52'$	14.7	SO	I1696	0-4-122/N530
27	22.3	$-01^o45'$	15.4	SBO		0-4-121?
29	22.3	$-01^o56'$	15.5	SO		0-4-120?
72	22.5	$+00^o10'$	14.9	SBO		0-4-125/I1697
8	22.8	$-01^o46'$	14.8	SBOp		0-4-129?
6	22.9	$-01^o48'$	14.7	Sbc	N530	0-4-130/N538
7	23.0	$-01^o45'$	14.8	SO		
4	23.0	$-01^o39'$	14.9	SO.	N535	0-4133/N535
33	23.0	$-02^o03'$	15.8	SO		

ZH	RA	Dec.	Mp	Type	NGC/IC	MCG
33	23.0	-02°03'	15.8	SO		
48	23.2	-01°44'	15.8	E3		
3	23.2	-01°37'	14.0	SO	NS41	0-4-157/N541
9	23.2	-01°35'	15.2	ES		
12	23.2	-01°42'	14.9	E4		
5	23.2	-01°34'	15.2	SBO		0-4-139?
10	23.3	-01°32'	15.0	E6		0-4-138/N543?
1	23.4	-01°35'	13.7	SO	N545	
2	23.4	-01°35'	13.4	E4/SO	N547	
11	23.5	-01°29'	15.1	SO	N548	0-4-141/N548
34	23.7	-02°02'	15.6	Sa?		
17	23.9	-01°53'	14.9	SBO	I1703	0-4-144/I1703
38	24.0	-01°21'	15.6	SBO		0-4-145?
40	24.1	-01°33'	15.6	Sb?		0-4-147?
37	24.2	-01°13'	15.2	Sa		0-4-148
51	24.7	-02°16	15.1	ES?	N558	
39	24.7	-01°31'	14.9	E4/SO		0-4-150
20	24.9	-02°09'	14.2	SO	NS60	0-4-151
19	25.0	-01°20'	14.8	SO		0-4-152?
41	25.1	-01°22'	15.1	SO		0-4-133?
21	25.3	-02°07'	14.1	E2/SO	N564	0-4-154/N564
26	25.4	-02°17'	15.0	SBO/a	I119	0-4-157/I119
53	25.5	-01°00'	15.2	E2		
18	25.6	-01°32'	14.6	Sa	N565	0-4-158/N565?
46	25.7	-02°09'	15.3	ES	I120	
61	25.9	-02°51'	15.2	SBcd		
45	26.1	-01°58'	15.0	SBO/a		
25	26.4	-01°11'	14.3	SBa/b	N570	0-4-162/N570
54	26.4	-00°49'	14.5	SO/a		0-4-163?

ZH	RA	Dec.	Mp	Type	NGC/IC	MCG
64	27.2	-01°30'	14.9	SBO		0-4-164
62	27.2	-02°14'	15.4	SO		
63	28.1	-02°15'	14.3	SBb		0-4-165/N577?
65	28.6	-01°45'	14.8	SBO		0-4-167
66	29.1	-01°11'	14.4	Sb		0-5-1/N585?
70	30.4	-00°57'	14.9	Sc		0-5-3/I138?
67	31.0	-01°31'	15.2	Sb		0-5-7
68	31.5	-01°20'	15.0	SBc		0-5-9
69	31.6	-01°17'	14.7	SaB		0-5-10?

Comments on Identifications

In the above table MCG identifications are compared with the Zwicky-Humason identifications, and the former, where followed by a question mark, represent probable coincidences of identifications but having positions differing by ±1' of arc in either coordinate. These are very probably correct in uncrowded regions, when the Zwicky-Humason identification correspond well with plate material such as the plate in Shapley's 'Galaxies' 2nd Edition page 179, fig. 106. However, since the Zwicky-Humason data was based on large scale plates, whilst the MCG was based on the POSS, whose many faint galaxies present images almost indistinguishable from stars, the Z-H identifications are to be preferred. Note that many galaxies in the Z-H list do not appear in the NCG, the converse also applies, many MCG objects do not appear in the Z-H list. However, the above comment still applies.

The MCG↔Z-H relationship is very confused in the central region, as is the RNGC. The main confusion is over Z-H NGC.545. Refering to the original NGC, we find the following data:

NGC.545 1^h18'51" -2°4.0'(1860) Stellar P of D neb.

NGC.547 1^h18'53" -2°4.3' Stellar F of D neb.

These relative positions are in good agreement with plate material and coincide with the Z-H identifications. However the MCG gives

MCG 0-4-140 = NGC.545 $01^h23.3$ -01°35' (1950), 15 mag.

MCG 0-4-142 = NGC.547a $01^h23.5$ -01°37', 13.7 mag.

MCG 0-4-143 = NGC.547b $01^h23.5$ -01°37', 13.8 mag.

The MCG position for NGC.545 fits the galaxy ZH-5 marked on the chart which forms a triangle with a pair of faint stars P. It is unlikely that D'Arrest, the visual discoverer, would take this as the component of a double nebula when one of the grandest in the sky lies slightly following! Indeed, this is where visual observation has the edge over photographic examination, for 545/547 both have stellar nuclei. These make the double nebula very obvious visually, but are burnt out in plates. The RNGC makes the same mistake. It quotes:-

RNGC 545 $01^h24.7'$, -01°28' (1975), 13.5 mag.
E,R,ALMSTEL,2*NR

RNGC 547 $01^h24.8'$, -01°28', 13.5 mag.
E,R,BM,CONT W/547A

RNGC 547A $01^h24.8'$, -01°28', no mag. given
E,R,BM,CONT W/547

They again identify NGC.545 with ZH-5, but quote a ludicrously bright magnitude - the object is very difficult in a 16 inch reflector.

There are the following ambiguities with regard to other bright galaxies.

(a) IC/1696/NGC.530 The RNGC quotes NGC.530 = IC1696, but also gives the Dreyer description EF,S,ME,F*SF. This fits in well with the nebula thus marked on the chart, ZH 13, which Z-H identify with IC106. The 1860 positions of IC106 and NGC.530 do not coincide, neither do the descriptions, the faint star SF mentioned by Dreyer being prominent visually. It would appear that NGC.530 = ZH 13 not ZH 6 which has a faint star NP. Also Z-H identify IC1696 as ZH 24 just SF ZH 13 where a round nebula appears on the plate. This accords with Dreyer -IC1696: EF, ES, 530 NP. Hence RNGC530 = IC1696 is clearly wrong, whilst IC106 is still ambiguous.

(b) NGC.538/ZH6 At the position of ZH6 on the plate, there is an oblique spiral galaxy with a faint star NP, this is identified with NGC.530 which, we recall, has a star SP. But ZH6 = MCG 0-4-130 = NGC.538. The MCG identification is the correct one here, since the original NGC description is NGC.538 EF, S, ME, F*N.

(c) IC1703/RNGC557 Zwicky and Humason identify ZH17 = IC1703. The former is shown as a basically round object with diffuse wings on the plate. The RNGC sites a galaxy NGC.557 at this position, with Dreyer description EF, S, R*10 NF, whilst the revised description is E,R,BM,IN CL, which adds little. The IC description for IC1703 is EF,S,DIF. It is therefore possible that ZH 17 = IC1703 = MCG 0-4-144 = NGC.557.

(d) NGC.543/ZH5 The position of NGC.543 relative to 545/547 quoted in the NGC and the RNGC new description, SLEL,BM,COM$ 2*CLOSES yield ZH 5 = NGC.543 = MCG 0-4-139.

Visual Observations

Four cluster fields containing 12 galaxies were observed using a 16 inch Newtonian. Descriptions of these galaxies in order of RA follow.

NGC.530 Small and slightly elongated with PA of approximately 45°. About 1' × ½'. Quite bright and brighter to the middle. Star off SF tip.

ZH 27 = MCG 0-4-121 Galaxy north following N530, not spotted on first observation but obvious to direct vision once spotted in the same MP field as NGC.530. Small and slightly elongated approximately N/S.

ZH 8 = MCG 0-4-129 First of 3 galaxies in a MP field. Small, slightly elongated approximately N/S. Faint and slowly brighter to the middle.

NGC.538 Brightest of 3. Elongated approximately N/S. Quite bright and brighter to the middle. Star on NF tip.

ZH 7 (not in MCG) 3rd of 3. Small and slightly elongated approximately E/W. Faint but brighter to the middle. Two faint stars in line close F.

NGC.535 Small and slightly elongated approximately E/W. Quite bright and slowly brighter to the middle. First of 5 galaxies in a MP field.

NGC.541 Small, (about 45" in diameter) and round. Quite bright and brighter to the middle.

NGC.543 Small, almost stellar and faint. Easy to direct once picked up.

NGC.545/547 Very impressive object. Pear shaped halo oriented approximately E/W with the wide end (NGC.545) preceding. The halo brightens to the middle considerably with two conspicuous stellar nuclei. Bright overall and about 1' major axis.

Abell 194

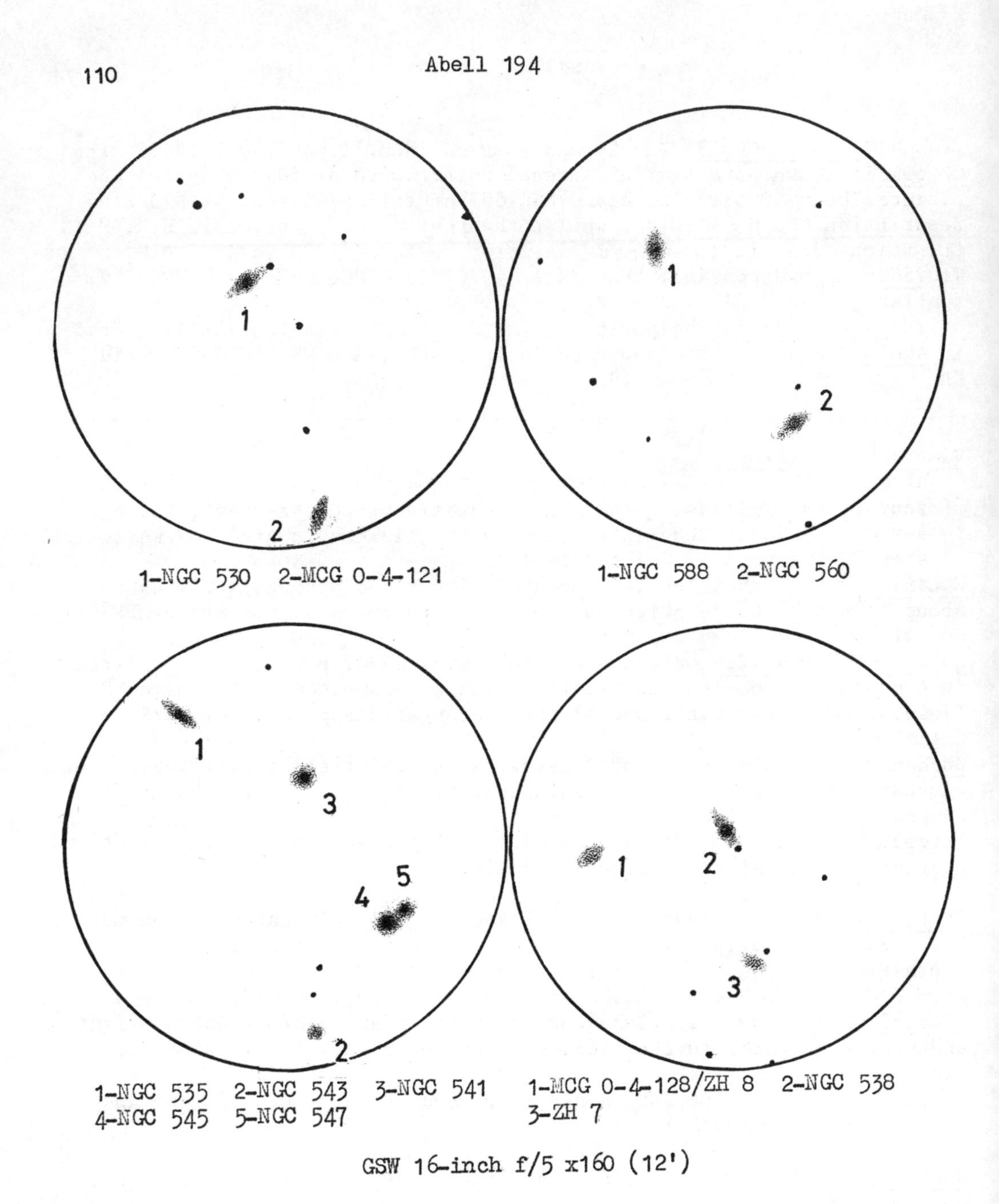

1-NGC 530 2-MCG 0-4-121

1-NGC 588 2-NGC 560

1-NGC 535 2-NGC 543 3-NGC 541
4-NGC 545 5-NGC 547

1-MCG 0-4-128/ZH 8 2-NGC 538
3-ZH 7

GSW 16-inch f/5 x160 (12')

NGC.568 First of two galaxies in a MP field. Quite bright and much brighter to the middle to a stellar nucleus. Slightly elongated N/S about 1' × ½'.

NGC.560 North following NGC.558. Similar to NGC.558 overall but no stellar nucleus, PA about 45°.

3. Abell 262

Abell	RA	Dec.	AD	AR	Z	RA
194	$01^h49.9'$	$+35^o55'$	1	0	0.0167	I

Although Abell 262 has Abell richness class 0, like Abell 194, its closeness (100 Mpc) and hence the brightness of its galaxies, makes it a visually rewarding cluster for moderate amateur apertures. The cluster has been shown to be a condensation in a supercluster of galaxies which stretches over about 10° of the sky through the NGC.507 group to the Pisces group. Both of the latter groups are rich visually and will be described separately in the section on groups of galaxies.

A glance at the chart of Abell 262 shows it to be a relatively loose, irregular cluster with a large proportion of spiral galaxies relative to ellipticals and lenticular galaxies. Recent work by Dickens and Moss(1) has shown that the cluster has a tight core of E and SO galaxies with an extended halo of spiral galaxies. They also showed that the velocity dispersion for the halo spirals is significantly larger than that of the E and SO galaxies, this may be due to a continuing gradual collapse of the halo.

List of Catalogued Galaxies in Abell 262

The following data is mainly taken from the MCG and supplemented from Dickens and Moss. It contains the CGCG number, position (1950.0). morphological type, NGC/IC number and MCG number. The latter are used on the chart of the central region which because of its large scale, also serves as a finder chart. Types are due to Dickens and Moss or guessed by the author from a 16" × 20" copy of a 48" Schmidt photograph.

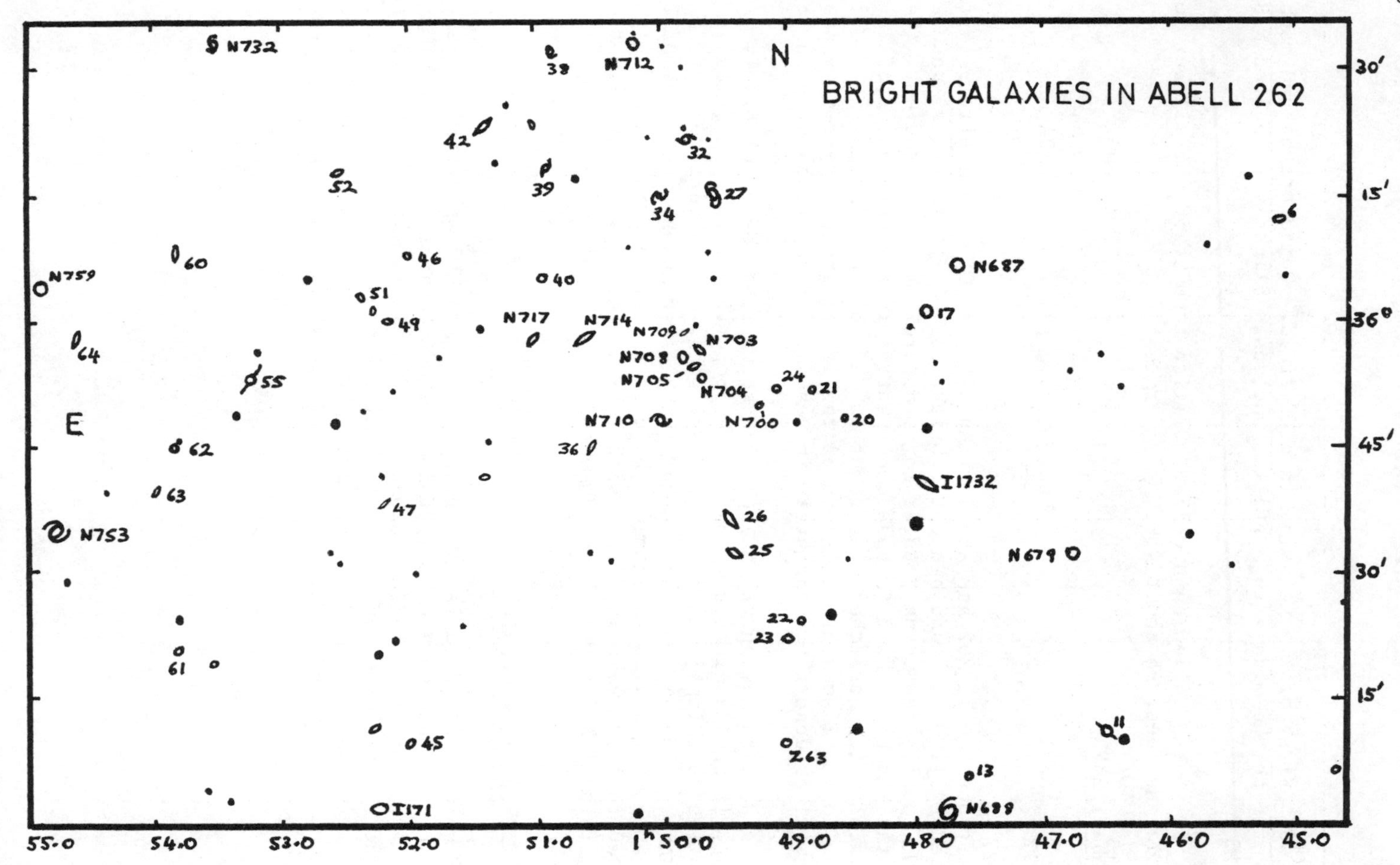

BRIGHT GALAXIES IN ABELL 262
N
E
30'
15'
36°
45'
30'
15'
55.0
54.0
53.0
52.0
51.0
1h50.0
49.0
48.0
47.0
46.0
45.0
N732
N712
N759
N687
N717
N714
N709
N703
N708
N705
N704
N710
N700
I1732
N753
N679
I171
N689

CGCG	RA	Dec.	Mp	Type	NGC/IC	MCG
25	01^h44.4'	35°17'	12.9	S	NGC.669	6-5-4/N669
	45.2'	36°12'	15	S		6-5-6
26	46.5'	35°11'	11	14.5	Sb	6-5-11
15	46.8'	35°32'	13.1	E	NGC.679	6-5-12
	47.6'	35°06'	15	I?Pec		6-5-13
12	47.65'	36°07'	13.3	E		6-5-14/N687
27	47.8'	35°01'	13.3	SBc		6-5-15/N688
	47.9'	35°40'	13	S		6-5-16/I1732
13	47.9'	36°01'	13			6-5-17
	48.55'	35°48'	13	S		6-5-20
	48.8'	35°52'	15	S		6-5-21
	48.9'	35°24'	15	S		6-5-22
	48.0'	35°22'	16	S		6-5-23
	49.1'	35°52'	14.8			6-5-24/N700
63	49.2'	35°10'	15.1	SO		
	49.45'	35°32'	14	S		6-5-25
	49.5'	35°36'	14.6	SB		6-5-26
10	49.6'	36°15'	13	SBc		6-5-27
4	49.7'	35°53'	14.1	E	NGC.704	6-5-28/N704
2	49.7'	35°55'	14.5	S	NGC.703	6-5-29/N703
3	49.75'	35°54'	14.5	SO	NGC.705	6-5-30/N705
1	49.8'	35°54'	14.8	E	NGC.708	6-5-31/N708
11	49.8'	36°22'	13.9	Sc		6-5-32
68	49.9'	35°59'	15.2'	E	NGC.709	not in MCG
	50.0'	35°48'	13	Sc		6-5-33/N710
	50.0'	36°15'	14	SBc(r)		6-5-34
22	50.2'	36°33'	13.9	S?E	NGC.712	6-5-35/N712
46	50.45'	36°42'	14.4	E		6-5-36
6	50.6	35°58'	13.9	SO	NGC.714	6-5-37/N714

CGCG	RA	Dec.	Mp	Type	NGC/IC	MCG
	$01^h50.85'$	$36^o32'$	14	S		6-5-38
	50.9'	$36^o19'$	15	Sc		6-5-39
8	50.9'	$36^o05'$	15.3	Sb		6-5-40
7	51.0'	$35^o58'$	14.7	S/SO	NGC.717	6-5-41/N717
	51.4'	$36^o23'$	14	S		6-5-42
	52.0'	$35^o10'$	15	S		6-5-45
	52.0'	$36^o08'$	16	S?		6-5-46
	52.0'	$36^o37'$	14	S?		6-5-47
32	52.05'	$36^o40'$	14			6-5-48
	52.2'	$36^o00'$	15	E		6-5-49
28	52.25'	$35^o02'$	13.8	Sb/E?	IC171	6-5-50/I171
	52.4'	$36^o03'$	15	S		6-5-51
	52.55'	$36^o18'$	16	S		6-5-52
	53.2'	$35^o53'$	13	Sc		6-5-55
45	53.5'	$36^o33'$	15	SB	NGC.732	6-5-57/N732
3P	53.8'	$36^o08'$	14.5	S		6-5-60
	53.8'	$35^o20'$	14	S		6-5-61
	53.85'	$35^o45'$	16	2E		6-5-62
	54.0'	$35^o41'$	15	S		6-5-63
	54.6'	$35^o58'$	15	S		6-5-64
33	54.8'	$35^o40'$	12.6	Sc	NGC.753	6-5-66/N753
34	54.85'	$36^o06'$	13.7	E	NGC.759	6-5-67/N759

Comments on Identifications

NGC.700 In the original NGC, Dreyer describes NGC.700 as VS,R,SP703. The CGCG and MCG positions differ however:- CGCG $01^h49.2'$, $35^o50'$, MCG $01^h49.1$ $35^o52'$. To add to the confusion, examination of a 16" × 20" photograph of A262 taken with the Palomar 48" Schmidt, reveals 5 galaxies which could conceivably fit the NGC description.

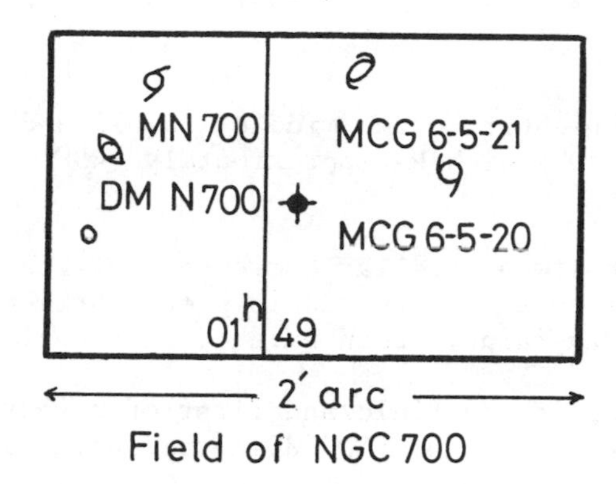

Field of NGC 700

It would appear from examination of the plate that Dickens and Moss have the correct identification in that DMN700 is a high surface brightness object, possibly an edge-on SO galaxy, whilst MN700 is a low surface brightness object, possibly a face-on Sc galaxy. With apparent magnitude less than 15, it is probably almost impossible to observe visually in amateur instruments.

NGC.709 Dickens and Moss identify a small elliptical galaxy north following the NGC.703/704/705/708 group as NGC.709. No galaxy appears in the MCG at this position. However, the original NGC description VF,PS,BET2*,GROUP SP is a very good description of the object on the plate. The RNGC description, SLEL,ALMSTELL also tallies well. Hence NGC.709 is wrongly omitted from the MCG.

Visual Observations

Abell 262 has been visually observed with the 36 inch Macdonald reflector in the USA, by R. Buta who presented a detailed drawing of the immediate surroundings of the NGC.703-708 group. A wider area of the cluster has been surveyed using a 16" reflector and 14 galaxies were picked up. Of these, 9 are NGC galaxies and 5 appear only in the MCG.

NGC.679
(16) Quite bright and slightly elongated N/S with dimensions of about $1\frac{1}{2}$' × 1'. Much brighter to the middle to a faint stellar nucleus.

NGC.687
(16) Quite bright and much brighter to the middle. Round, with a diameter of about 1'. MCG 6-5-17 SF was not spotted.

MCG 6-5-26

(16) Quite bright and brighter to the middle. Small and slightly elongated (about 45" × 30") with PA approximately -45°.

NGC.703

(36) Double nebula with almost stellar components aligned NS, the brighter component lies to the north and is pretty bright, the companion being described as pretty faint. Both have stellar nuclei.

(16) First of 5 galaxies in the field and first of a very compact group of four galaxies. Small and round (30" diameter) being quite bright and brighter to the middle.

NGC.704

(36) Round, considerably bright and much brighter to the middle with a bright nucleus. Two very faint stars close SP.

(16) Second of a compact group of 4 and most southerly. Similar to 703 but fainter.

NGC.(705)

(36) Pretty bright and brighter to the middle to a stellar nucleus. Much extended NP/SF. Very faint almost stellar galaxy close preceding.

(16) Faint and small. Two stars very close south. Third and faintest of the compact group.

NGC.708

(36) Considerably bright and brighter to the middle. Round with a faint star close north and an extremely faint almost stellar galaxy following.

(16) Fourth of the compact group of 4 and brightest and longest. Round with a diameter of about 1'. Quite bright and brighter to the middle.

MCG 6-5-27

(16) Quite large and diffuse oval with low surface brightness. First of two galaxies.

MCG 6-5-32

(16) Quite bright. Very small, being less than 30" along the major axis. Spindle very suddenly much brighter to the middle with a high surface brightness core. Slightly elongated N/S. Second of two galaxies.

Abell 262

GSW 16-inch f/5 x160

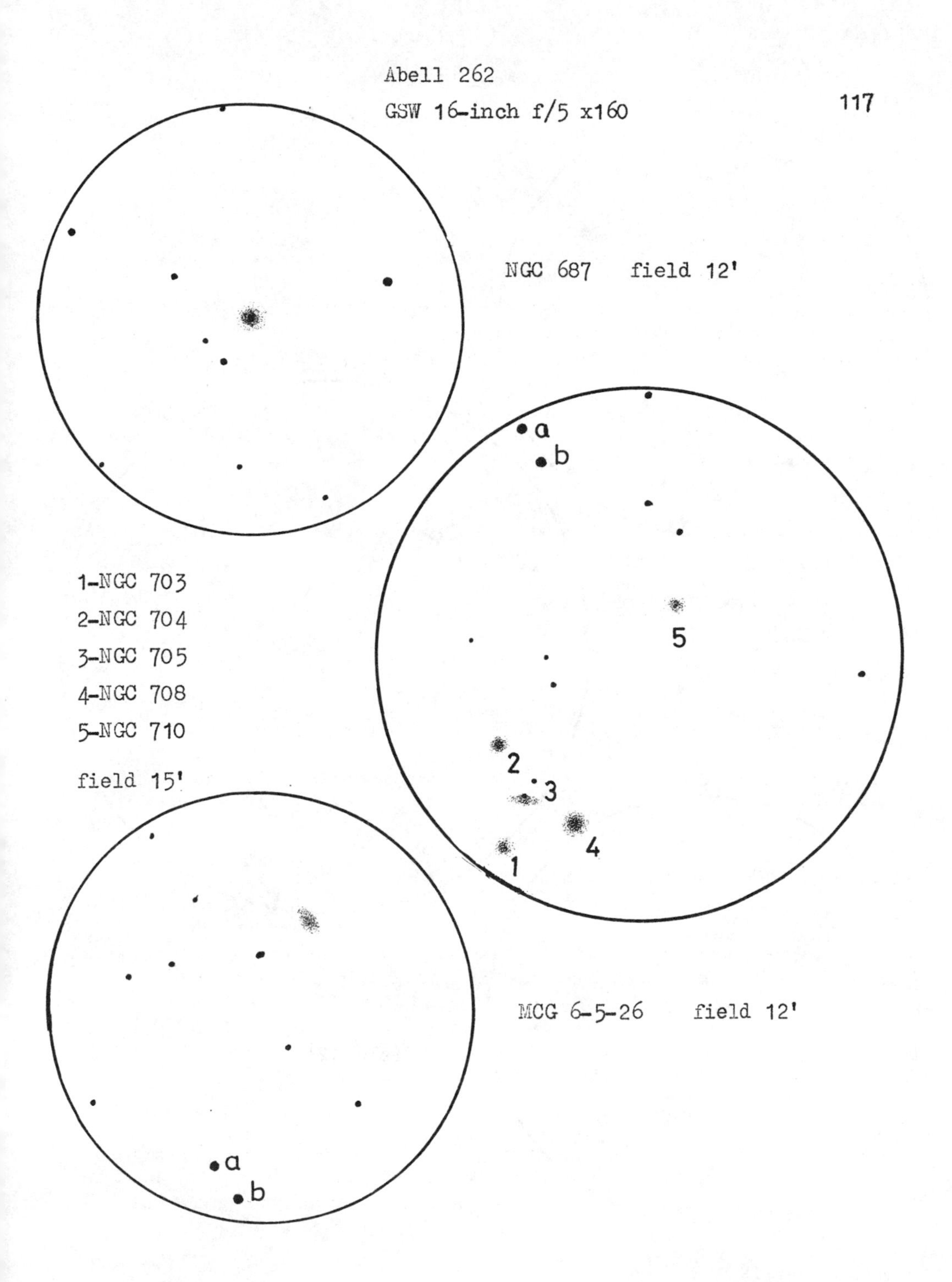

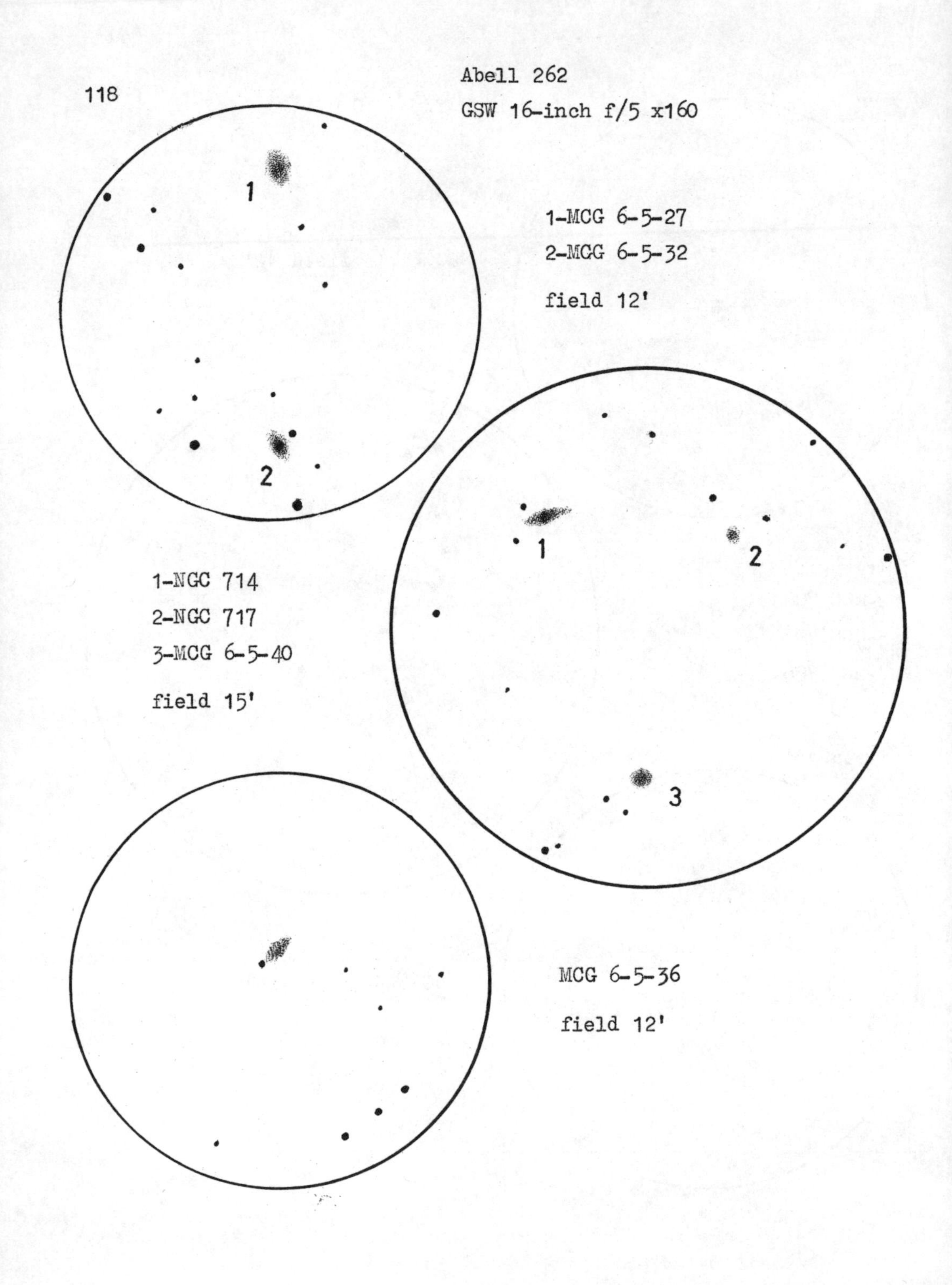
Abell 262
GSW 16-inch f/5 x160
1
2
1-MCG 6-5-27
2-MCG 6-5-32
field 12'
1
2
3
1-NGC 714
2-NGC 717
3-MCG 6-5-40
field 15'
MCG 6-5-36
field 12'

MCG 6-5-36
(16) Quite bright and slowly brighter to the middle. Elongated approximately N/S and having a star of magnitude 11 or 12 on the N tip.

NGC.709
(36) Pretty faint nebula much extended NP/SF with a bright nucleus.

NGC.710
(36) Pretty faint, diffuse nebula which is considerably large and brightens considerably to the middle. Slightly elongated N/S.

(16) Quite bright and a little brighter to the middle. Slightly elongated EW with a diameter of about 45". Last of a field of five galaxies.

NGC.714
(16) First of a field of 3 galaxies. One of the easiest objects after NGC.708 group. Quite bright and brighter to the middle. Spindle elongated approximately E/W with two faint stars close proceeding.

MCG 6-5-40
(16) Second of a field of three galaxies. Round with a diameter of about 30". Quite bright and brighter to the middle, but inferior to NGC.714. Group of stars of mag. 12 proceding.

NGC.717
(16) Small, irregularly round and faint. Difficult. Third of 3.

4. Abell 347

Abell	RA	Dec.	AD	AR		RS
347	$02^h22'$	$41^o35'$	1	0	0.019	I?

Abell 347 is at the end of a supercluster starting at A426, the Perseus cluster, which has a similar redshift. The cluster is rather poor in galaxies although it presents an interesting visual spectacle. The radio source 3C66 has been shown to be associated with a cluster galaxy(1) and shows a head-tail structure. The velocity dispersion due to galactic orbital motions is small and this is consistent with the dependence of X-ray flux on the velocity dispersion.

List of Catalogued Galaxies in Abell 347

The following table is based upon the list of galaxies published in a paper by Hintzen et al(1). These are cross correlated with the MCG which also supplies photographic magnitudes. Hubble types for some of the brightest galaxies have been estimated from a polaroid elargement

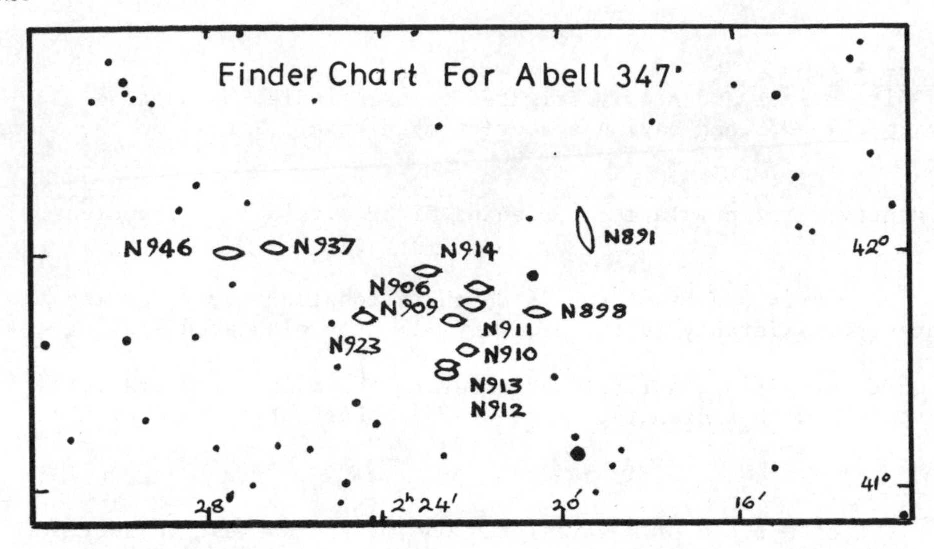
Finder Chart For Abell 347
N946
N937
N914
N891
N906
N909
N898
N911
N923
N910
N913
N912
42°
41°
28
2h 24'
20'
16'

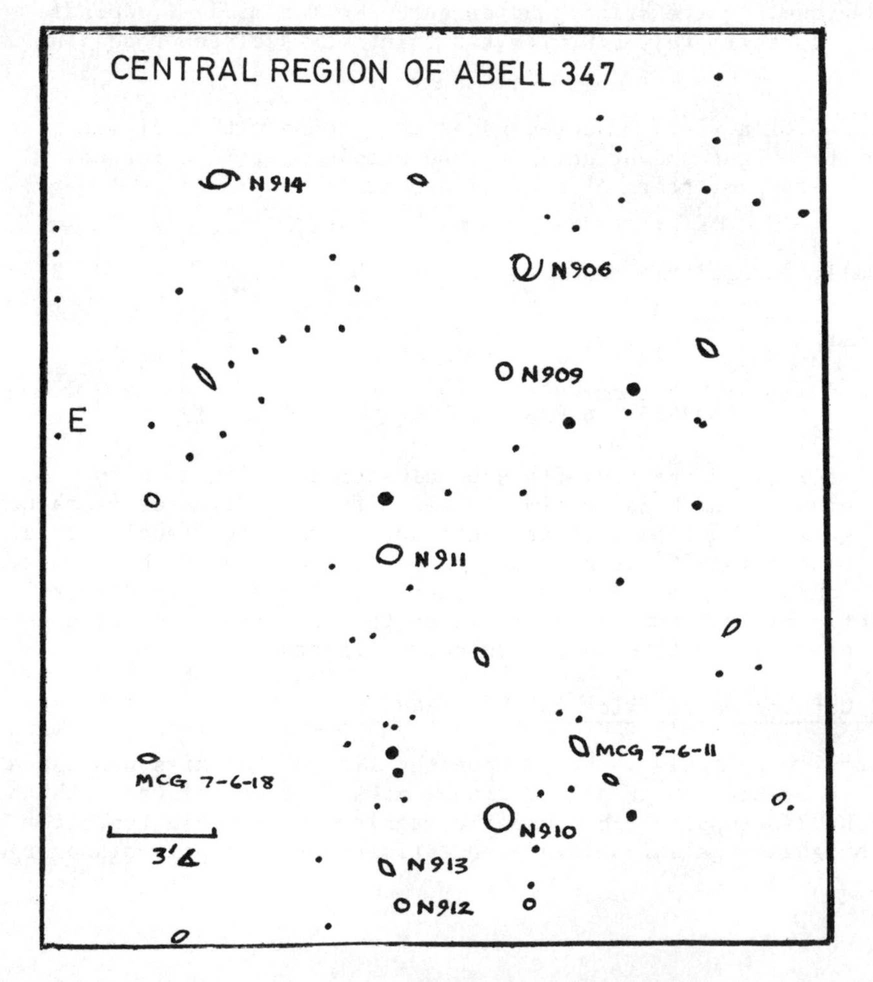
CENTRAL REGION OF ABELL 347
N914
N906
N909
E
N911
MCG 7-6-11
MCG 7-6-18
N910
3'6
N913
N912

of part of a POSS plate which was also used to draw the chart of the central region.

CGCG	RA	Dec.	Mp	Type	NCG/IC	MCG
1	$01^h58'24''$	$44^o46'0$	15		background	7-5-5
2	$02^h02'54''$	$44^o58'$	14			7-5-12
3	03'42"	$44^o20'$	13		NGC.812	7-5-14/N812
4	24'24"	$41^o45'$	14		NGC.923	7-6-22/N923
5	20'00"	$42^o26'$			3C66	
6	23'36"	$41^o37'$	15			7-6-20
7	29'12"	$41^o59'$	15		background	7-6-30
8	22'54"	$41^o55'$	14	Sa/b	NGC.914	7-6-17/N914
9	22'36"	$41^o45'$	15	E?	NGC.911	7-6-16/N911
10	21'36"	$42^o24'$	15			7-6-9
11	20'12"	$41^o44'$	14	SO	NGC.898	7-6-4/N898
12	27'30"	$42^o01'$	15		NGC.946	7-6-26/N946
13	22'06"	$41^o52'$	14	SBb	NGC.906	7-6-12/N906
14	22'18"	$41^o36'$	14	E	NGC.910	7-6-14/N910
15	21'42"	$43^o06'$				
16	22'12"	$41^o49'$	15	E?	NGC.909	7-6-13/N909
17	19'48"	$42^o47'$				
18	28'06"	$43^o15'$				
19	29'00"	$43^o14'$				7-6-29?
20	24'12"	$40^o50'$				

Not all the MCG galaxies in the area covered by the above list are included. However all the MCG galaxies on the identification chart are marked in.

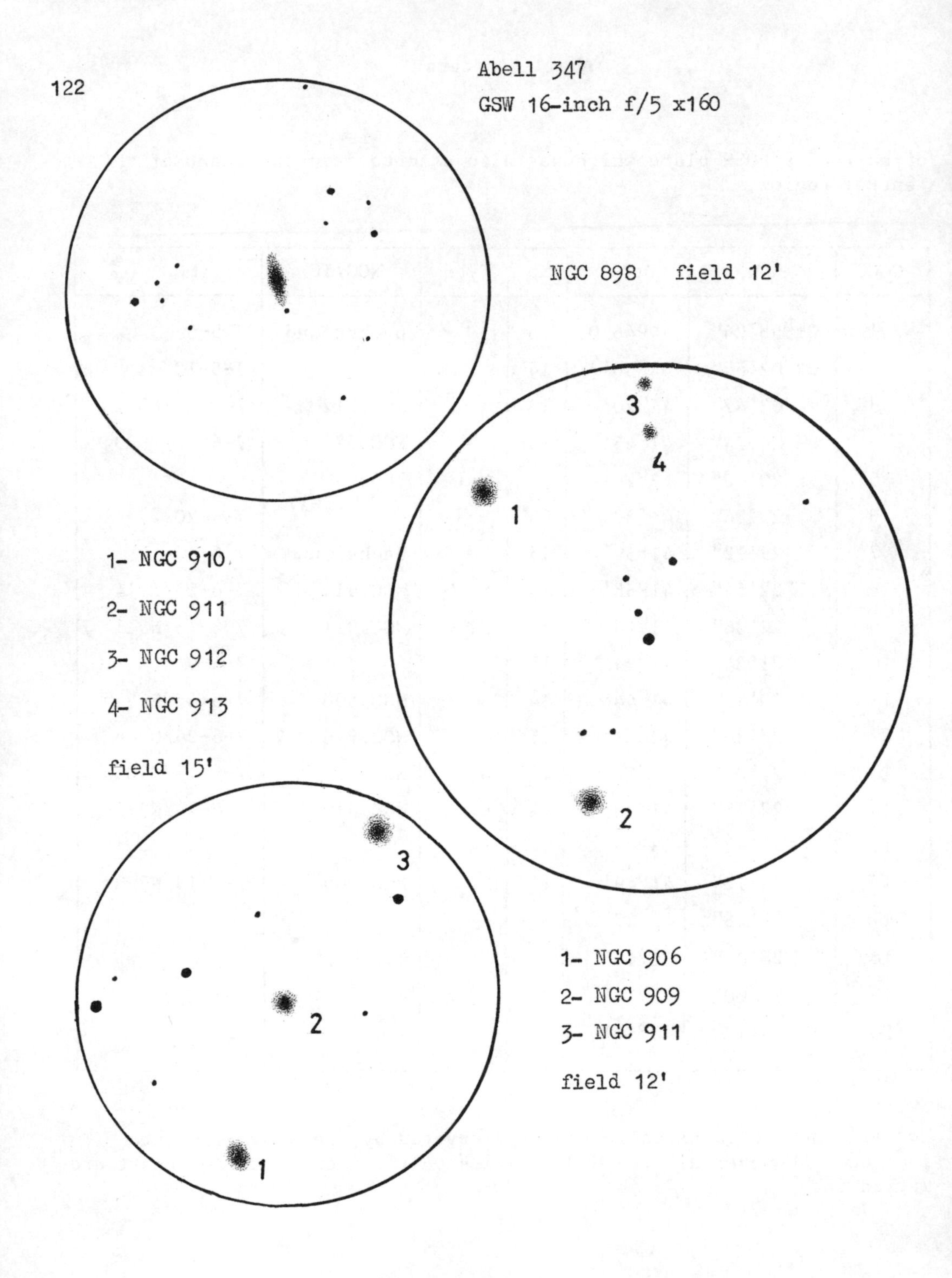
Abell 347
GSW 16-inch f/5 x160
NGC 898 field 12'
3
4
1
2
1- NGC 910
2- NGC 911
3- NGC 912
4- NGC 913
field 15'
3
2
1
1- NGC 906
2- NGC 909
3- NGC 911
field 12'

Visual Observations

Because A347 lies about $\frac{1}{2}^{o}$ south preceding the magnificent, bright edge-on foreground galaxy NGC.891 which is itself a member of the nearby NGC.1023 group, it is easy to find and once found, offers some rewarding visual galaxy fields. The following galaxies were observed using a 16 inch f/5 Newtonian reflector.

NGC.898 Quite bright and much brighter to the middle. Spindle elongated approximately N/S. Almost in the same low power field as NGC.891. About 1' × 20" with a faint star near northern tip.

NGC.906 Faint and slowly brighter to the middle with low surface brightness. Slightly elongated approximately N/S. First of 3 galaxies in a MP field. (1' × 40")

NGC.909 Quite bright and brighter to the middle. Round and about 45" in diameter. Second of three.

NGC.911 Brightest galaxy observed in the cluster. Round and quite bright brightening considerably to the centre. About 1' in diameter and third of a field of three. Also second of a field of four:-

NGC.910 First of a field of four galaxies. Similar to NGC.911 which is about 10'N, but slightly inferior in size and brightness.

NGC.912 Third of three. Faint and almost stellar. Close north of NGC.913.

NGC.913 Fourth of a field of four galaxies. Similar to NGC.912.

5. Abell 426

Abell	RA	Dec.	AD	AR	Z	RS
426	$03^h16.4'$	$41^o19'$	0	2	0.018	L

The Perseus cluster, Abell 426[1] is the only richness class 2 cluster with distance class 0, and is in fact the closest of the nearby rich clusters of galaxies. Although in the cluster lies beyond a spiral arm of the Milky Way, by some good fortune, galactic absorption in its direction is quite low. The largest and brightest cluster galaxy NGC.1275 has apparent photographic magnitude 12.7, and is a very interesting object indeed. First of all it was on Seyfert's original list of emission line galaxies, and is therefore called a Seyfert galaxy. Seyfert galaxies have very bright nuclei which, as in the case of NGC.1275, are of relatively rapidly varying brightness, showing a system of emission lines in their spectra. NGC.1275 is unique in being the only E galaxy in the class, and it has been proposed that the galaxy may in fact be a BL-Lacerta object[1]. The latter class of object are usually almost stellar, having characteristic radio and optical spectra devoid of emission lines. A few of them have been shown to be embedded in the cores of elliptical galaxies.

Part of the reason for placing the emission line galaxy NGC.1275 in the BL-Lacerta category is that NGC.1275 would be by far the closest object of its kind. The galaxy itself also has a double redshift system, with a low redshift system associated with most of the galaxy and the nucleus, and a high redshift system associated with a clump of knots to the north of the galaxy. Recent work[3], [4] has pointed to the strong possibility that the high redshift system is caused by a rapidly moving, early type spiral galaxy that happens to lie in the line of sight to NGC.1275. Indeed, plates of NGC.1275 taken through a filter tuned to the low redshift system show the galaxy to be a perfectly symmetrical EO. Another interesting feature of this extraordinary object is a system of faint filaments shining in Hα which apparently surge out of its nucleus. There are two current explanations for these. The earliest is that they could have originated on some explosion in the undoubtedly violently active nucleus of the galaxy - similar to the 'Crab' Nebula, but on a far grander scale. A more recent explanation is that the filaments could be condensations in an accretion flow into the nucleus from the intracluster medium.

The cluster as a whole is an X-ray source[5] superimposed on which is the source associated with the nucleus of NGC.1275. Ryle and Windram claimed to have detected a diffuse cluster radio source, but this has not been confirmed by more recent observations[6]. The cluster X-ray source is associated with the intracluster medium, which has produced spectacular head-tail morphologies in cluster radio galaxies, amongst which is the classic of its class - NGC.1265. NGC.1275 is also a radio galaxy but appears symmetric to most radio telescopes. Intercontinental interferometry[7] has shown the presence of a tiny double structure with jets around the nucleus.

As a whole, the cluster is highly elongated along a line from NGC.1275 to IC310, around which are distributed most of the bright galaxies. The cluster X-ray source is elongated along this line, and has been successfully modelled as being due to heat radiation from a prolate ellipsoidal distribution of gas along the chain. The proportion of ellipticals and SO galaxies to spirals is high, indicating a relatively evolved cluster, most of the spirals having been stripped of gas in their passage through the core. The lack of spherical symmetry seems at variance with the degree of evolution and remains unexplained.

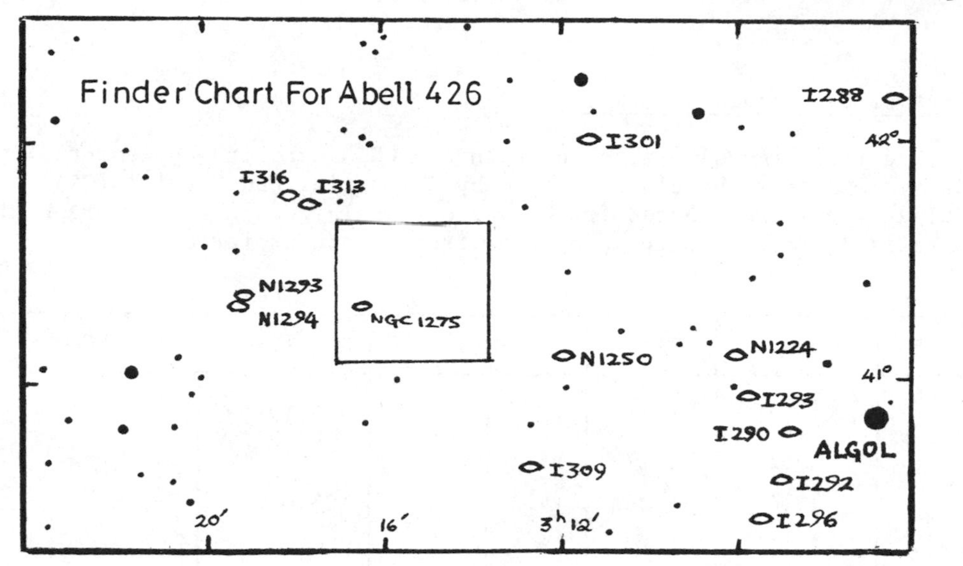
Finder Chart For Abell 426
I288
I301
42°
I316
I313
N1293
N1294
NGC 1275
N1250
N1224
41°
I293
I290
ALGOL
I309
I292
I296
20′
16′
3h 12′

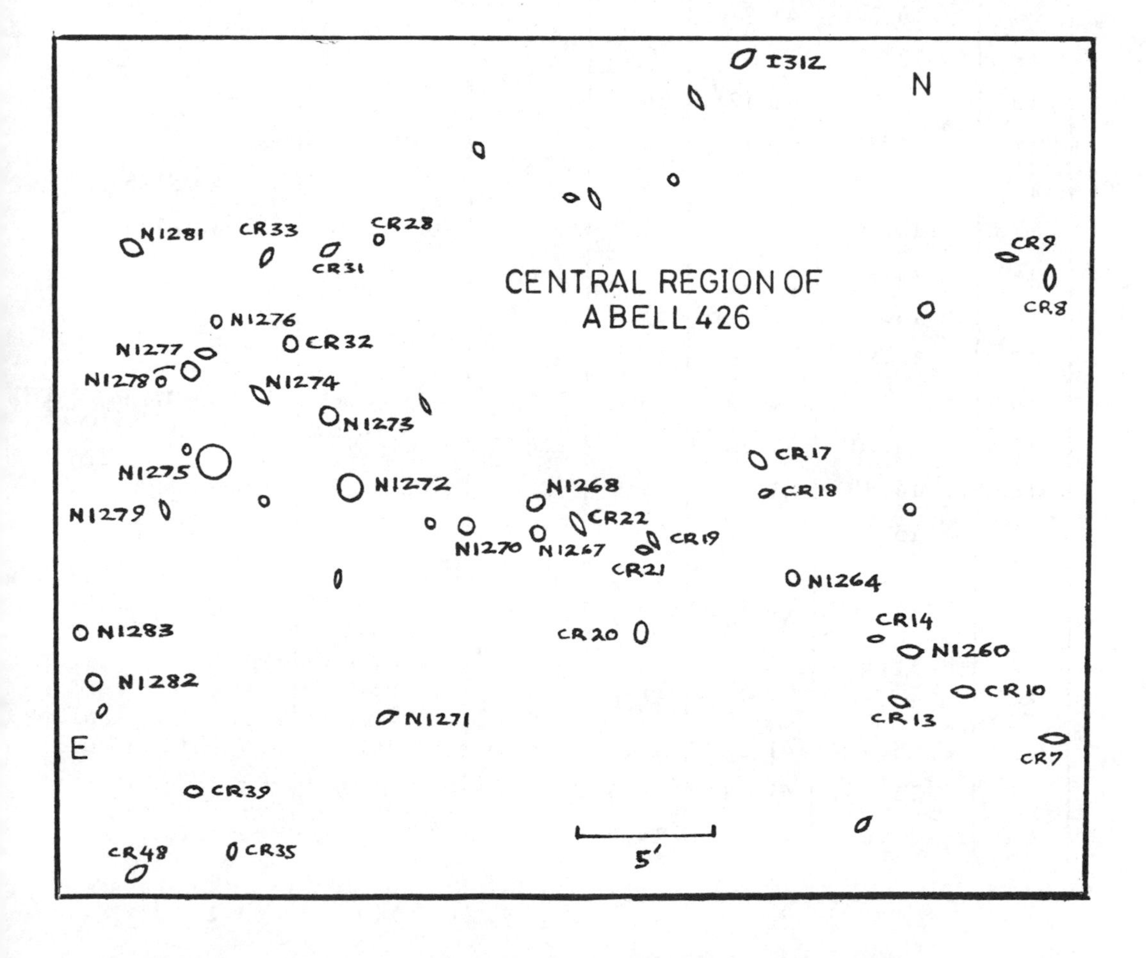
I312
N
N1281
CR33
CR28
CR31
CR9
CR8
CENTRAL REGION OF
ABELL 426
N1276
CR32
N1277
N1278
N1274
N1273
CR17
CR18
N1275
N1272
N1268
N1279
CR22
CR19
N1270
N1267
CR21
N1264
CR14
N1283
CR20
N1260
N1282
CR10
N1271
CR13
E
CR7
CR39
CR48
CR35
5′

List of Catalogued Galaxies

The following data is based upon the CGCG, and is supplemented by data drawn from the classic paper by Chincarini and Rood[1] (CR). Galaxy types are guessed from a variety of large scale photographic material. Photographic magnitudes from CR are preferred.

CR	RA	Dec.	Mp	Type	NGC/IC	MCG
6	$03^h13.2'$	41 21	14.7		NGC.1257?	7-7-44
	13.2'	41 27	15.7			
	13.4'	41 08	14.3		IC310	7-7-45
7	13.7'	41 10	15.7			
8	13.7'	41 27	15.0			
9	13.9'	41 28				
10	14.0'	41 12	16			7-7-46
11	14.2'	41 18				
13?	14.2'	14 11	15			7-7-47/N1259
12	14.2'	41 13	14.2		NGC.1260	7-7-48/N1260
14	14.3'	41 13				
	14.5'	41 47	15.6			
15	14.6'	41 16	15.4			
16	14.65'	14 19	15			7-7-50/N1264
17	14.70'	41 20		Pec		
18	14.8'	41 34	14.9		IC312	7-7-51
19	15.0'	41 17				
20	15.1'	41 14	15.3			
	15.1'	41 40	14.7		NGC.1265	7-7-52/N1265
21	15.1'	41 17				
22	15.3'	41 17	15.7			
23	15.5'	41 17	15.4		NGC.1267	7-7-53/N1267
24	15.5'	41 18	14.5	Sc	NGC.1268	7-7-56/N1268
25	15.7'	41 17	14.4		NGC.1270	7-7-57/N1270

CR	RA	Dec.	Mp	Type	NGC/IC	MCG
26	$01^h15.9'$	41 10	15.4		NGC.1271	
28	16.0'	41 28				
29	16.1'	41 18	14.5		NGC.1272	7-7-58/N1272
30	16.2'	41 21	14		NGC.1273	7-7-59/N1273
31	16.2'	31 27	15.5			7-7-60
32	16.3'	41 24	15.2			7-7-61/IC1907
34	16.4'	41 22			IC1907 = NGC.1274	7-7-62/N1274
	16.5'	41 20	13.0	E0pec	NGC.1275	7-7-63/N1275
35	16.5'	41 06				
36	16.5'	41 25				
37	16.6'	41 23	14.9		NGC.1277	7-7-64/N1277
38	16.7'	41 22	14.4		NGC.1278	7-7-65/N1278
39	16.6'	41 07				
40	16.8'	41 04	15.6			
41	16.7'	41 18				
42	16.8'	41 27	15.0		NGC.1281	7-7-67/N1281
43	17.0'	41 11	14.3		NGC.1282	7-7-68/N1282
44	17.0'	41 13	15.6		NGC.1283	7-7-69/N1283

Comments on Identifications

The main comment is on the positional differences between the CGCG and the MCG, the Chincarini and Rood positions being based on the CGCG. In some cases the differences can lead to ambiguities in tight groups unless relative positions are used.

NGC.1257 This galaxy is not listed as such in the CGCG, the MCG or in CR. It is listed in the RNGC where the new description reads EL,BM,*CLOSE N. Plotting this object using RNGC positions for other cluster galaxies confirms this identification when reference is made to plate material. It is strange that although the star is rather bright, perhaps mag. 11 or 12, and the nebula is difficult (not found in a 16"), the original Dreyer description does not mention the star.

NGC.1264 The MCG names the galaxy CR 16 NCG.1264 if the positions are taken to be the same:- CR 16 has 1950 position $03^h14.7 + 41^o19'$ whilst MCG 7-7-50 or NGC.1264 has position $03^h14.65\ 41^o19'$. However, according to a Mayall 4 m plate used by Strom and Strom in(4), CR 16 is an extremely small virtually stellar object close south of CR 17, a relatively large and bright peculiar object. This has coordinates $03^h14.7$ $41\ 20^o$. It is therefore possible that MCG 7-7-50 = CR 17 not CR 16. In this case the identification NGC.1264 = MCGC 7-7-50 is more plausible. CR 16 is probably not visually observable in ordinary telescopes.

NGC.1279 This object is not named as such in the CGCG or MCG. According to the RNGC its position is 0.2' time east and 2' of arc S of NGC.1275. Its 1950 coordinates would therefore be $03^h16.7'\ 41^o18'$ which is not occupied by any CGCG or MCG object. However CR list galaxy no. 41 at this position. Thus CR 41 = NGC.1279.

NGC.1276 Again neither the CGCG or MCG lists any object with this identification. However, according to the RNGC, 1276 is 3' due north of NGC.1275. Its 1950 coordinates would therefore be $03^h16.5' + 41^o23'$. CR 36 is at $03^h16.5' + 41^o25'$, thus close NP NGC.1277 = CR 37. Inspecting plate material yields only one candidate: NGC.1276 = CR 36.

Visual Observations

Abell 426 is difficult to locate amongst the teeming throng of Milky Way stars without setting circles because most of its galaxies are of small apparent size and rather faint. However, NGC.1275 has been reported as visible in a 5 inch Celestron by Dave Branchett, and Ed Barker has seen a few of the brighter cluster members around NGC.1275 in an $8\frac{1}{2}$ inch Newtonian reflector. In larger apertures, the Perseus cluster presents probably the most spectacular concentrated galaxy field in the northern winter sky. The following observations were made using telescopes ranging in aperture from 36 inches to 5 inches. Most of the brighter galaxies are observable in a 16 inch reflector.

IC 310

(16) Small and round (about 25" in diameter). Faint, but slightly brighter to the middle.

NGC.1260

(16) Quite bright and slowly brighter to the middle. Elongated ellipse about 30" × 15" oriented roughly E/W. Triangle of stars of mag. 11 or 12 F.

NGC.1265

(16) Small and round (about 30" in diameter). Quite bright and brighter to the middle with a star of mag. 12 following. Very difficult to pick up amongst Milky Way stars.

NGC.1267

(16) First of a group of 5 galaxies in a 14' MP field. Small and almost stellar. Relatively high surface brightness - a fuzzy star.

NGC.1268

(16) Second of 5. Similar to NGC.1267 but less compact. Close F NGC.1267.

NGC.1270

(16) Third of 5. Similar to NGC.1268 but a little brighter.

NGC.1272

(36) One of a field of 16 galaxies in the central region around NGC.1275. Pretty faint diffuse round nebula slowly brighter to the middle.

(16½) In the field of NGC.1275. Resembles NGC.1275 in size and overall appearance but slightly smaller. Irregularly round and brighter to the middle.

(16) Fourth of a field of 5 starting near NGC.1267 being the largest and brightest in the field. Round (about 30" diameter) and quite bright and brighter to the middle. Also in a field of 9 galaxies around NGC.1275.

NGC.1273

(36) Small round nebula with a well defined nucleus. Considerably faint.

(16½) Small, irregularly round, and slightly brighter to the middle. In the field of NGC.1275.

(16) Fifth of 5 starting near NGC.1267, also in the field of NGC.1275. Quite bright and brighter to the middle. Round and about 25" in diameter.

(8½) Faint blur in the field of NGC.1275.

CR 28

(36) Small, round, faint nebula possibly having a stellar nucleus.

CR 31

(36) Small oval slightly extended NP/SF. Faint.

CR 32
(36) Small, round nebula. Considerably faint. North of a triangle of stars.

(16½) North preceding an equilateral triangle of stars in the field of NGC.1275. Small and irregularly round. Easy at medium power.

(16) Close north of a triangle of faint stars. Small and round. Difficult.

CR 33
(36) Very small, round nebula. Very faint.

NGC.1274
(36) Small and round. Considerably faint with a faint stellar nucleus.

(16) Small and round, perhaps 20" in diameter. Quite bright and brighter to the middle. Close NP NGC.1275.

NGC.1275
(36) Small and round, but the largest in the field. Pretty bright having a conspicuous stellar nucleus.

(16½) Fairly bright, small and round and brightening in the middle to a small stellar nucleus.

(16) Brightest and largest in a MP field of 9 galaxies. Quite bright and suddenly much brighter to a prominent stellar nucleus of about mag. 12.

(8½) Easy even at low power. Round nebula with a brilliant stellar nucleus.

NGC.1276
(36) Very small, round and considerably faint. An extremely faint, stellar galaxy close north following.

NGC.1277
(36) Very small and irregularly round. Considerably faint. Close NP NGC.1278.

(16½) Nebulous star-like object just NP NGC.1278.

(16) Quite bright and brighter to the middle. Very small but non-stellar. Close NP NGC.1278.

NGC.1278

(36) Pretty faint, small, round nebula slowly brightening to the middle. A very faint, very small round nebula lies close SP. Another SF.

(16½) Almost due north of NGC.1275. Small and irregularly round. Easy at medium power. A little brighter to the middle.

(16) Small and round. Superior in size and brightness to NGC.1277. Quite bright and brighter to the middle.

NGC.1279

(36) Considerably faint, small round nebula.

(16) Small object south of NGC.1275. Irregularly round. Quite bright and a little brighter to the middle.

CR 48

(16) First of a field of 3 galaxies containing NGC.1282/3. Small and round, about 20" in diameter. Quite bright and brighter to the middle. Two stars of mag. 12 NF.

NGC.1281

(16½) Fairly bright, medium size and irregularly round. Generally of uniform surface brightness but a little brighter in the centre. Star SP by about 45".

(16) Last of a field of 9 galaxies in the MP field of NGC.1275. Slightly elongated approximately E/W with a star close preceding. Quite bright.

(8½) Small, faint blur.

NGC.1282

(16) Second of 3 and largest and brightest. Close SP NGC.1283. Small and round (25"). Quite bright and brighter to the middle.

NGC.1283

(16) Third of 3. Close NF NGC.1282. Smaller and fainter than NGC.1282, being round, quite bright and brighter to the middle.

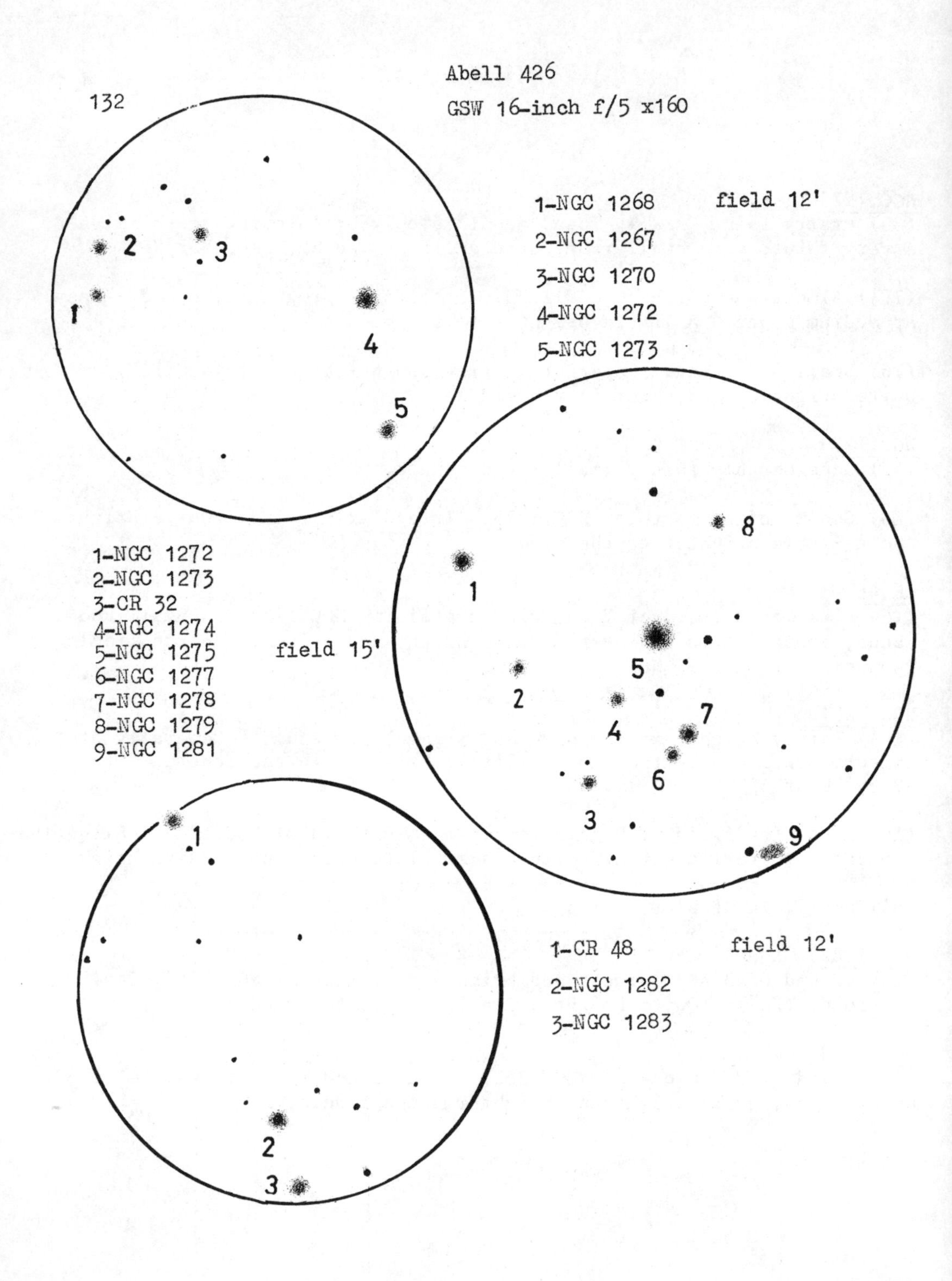
Abell 426
GSW 16-inch f/5 x160
1-NGC 1268
2-NGC 1267
3-NGC 1270
4-NGC 1272
5-NGC 1273
field 12'
1-NGC 1272
2-NGC 1273
3-CR 32
4-NGC 1274
5-NGC 1275
6-NGC 1277
7-NGC 1278
8-NGC 1279
9-NGC 1281
field 15'
1-CR 48
2-NGC 1282
3-NGC 1283
field 12'

6. Abell 1185

WS	Abell	RA	Dec.	AD	AR	Z	RS
	1185	$11^h08.2$	$+28^o57'$	2	1	0.035	C

Abell 1185 is an interesting cluster visually, containing six NGC galaxies in its core region. Its brightest galaxy is NGC.3550, and the cluster contains a very interesting Arp peculiar system - Arp 105. The latter consists of an E galaxy identified as NGC.3561 in the MCG, with a highly peculiar galaxy close north. The latter consists of two spirals in contact. This structure is noted by Vorontsov-Velyaminov, and is clear on a polaroid enlargement taken from the POSS. There is a reproduction of a 200 inch photograph in[1] which is overexposed to show faint details, and thus blurs out the double structure. The galaxies are clearly in tidal interaction, since plumes (visible even on the poor quality polaroid) spray out of the galaxies. The most prominant plume extends about 20" in PA -45° and has a peculiar galaxy apparently attached to one end. The SP galaxy has a counterplume extending towards NGC.3561 which also has a plume extending directly S away from the pair - MCG 5-27-11, which has yet another counterjet pointing due N away from NGC.3561. In[1], Arp points out that a quasar of redshift 2.19 is situated close preceding NGC.3561. This is probably coincidental, since the system of plumes and counterplumes is typical of mere tidal interaction - not uncommon in rich clusters. To add to the interest however, a dwarf galaxy known as 'Ambartsumian's Knot' is situated on the end of the NGC.3561 plume. The stalk and the knot is clearly visible on the POSS.

List of MCG Galaxies in the Central Region

MCG	RA	Dec.	Mp	NGC
5-27-2	$11^h08.0'$	$29^o02'$	14.2	3550
5-27-3	08.0'	$28^o58'$	15.1	3552
5-27-4	08.0'	$28^o59'$		3553
5-27-7	08.1'	$28^o56'$	15.3	3554
5-27-8	08.2'	$28^o49'$	14.8	3558
5-27-10	08.5'	$28^o58'$	14.7	3561
5-27-11	08.5'	$28^o59'$		
5-27-12	08.6'	$28^o58'$		
5.27-15	08.75'	$29^o00'$	15.5	
5-27-16	08.90'	$29^o01'$	15.7	

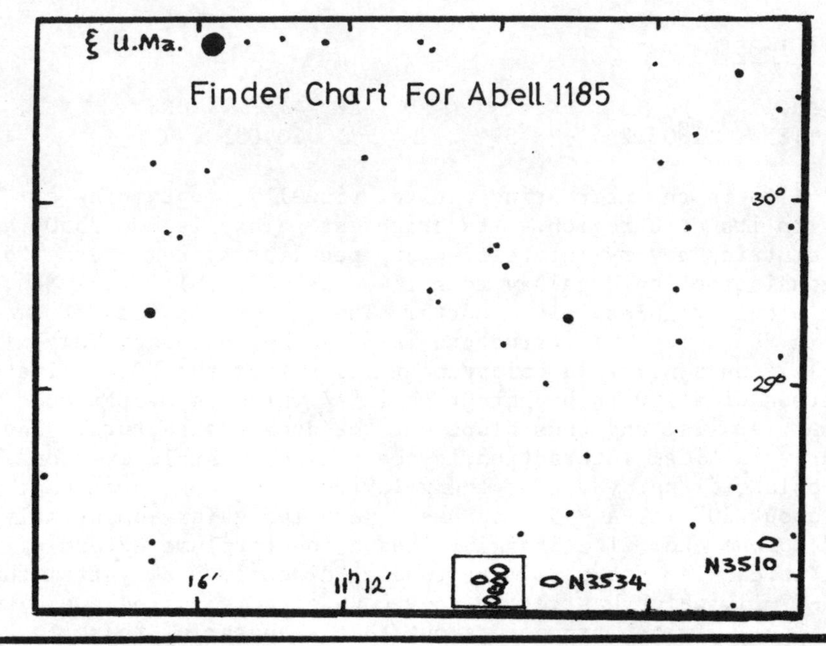
ξ U.Ma.
Finder Chart For Abell 1185
30°
29°
16′
11h 12′
N3534
N3510

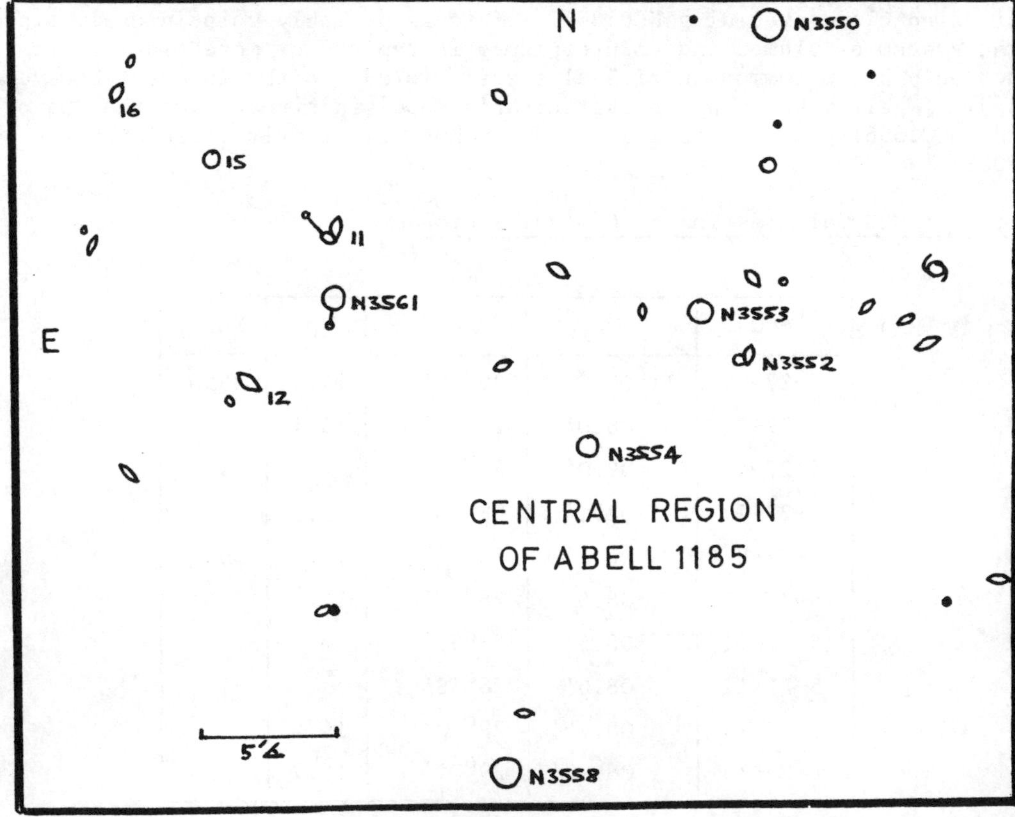
N
N3550
16
15
11
N3561
E
N3553
N3552
12
N3554
CENTRAL REGION
OF ABELL 1185
5′
N3558

Abell 1185

GSW 16-inch f/5 x160 (15')

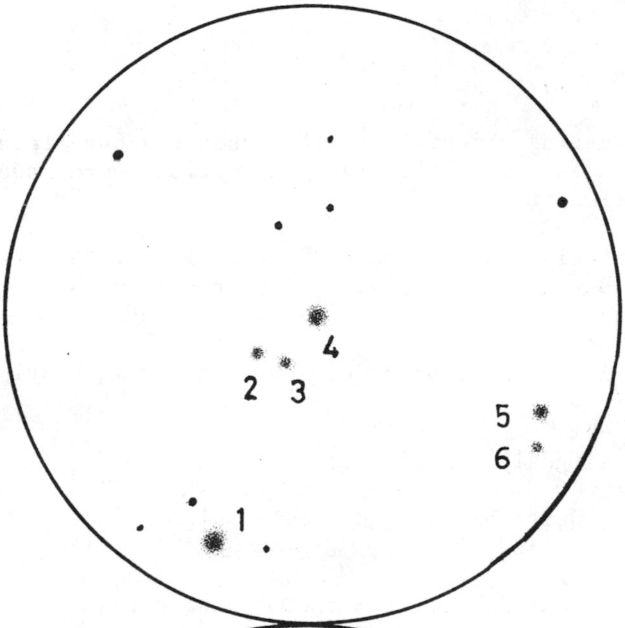

1-NGC 3550

2-NGC 3552

3-NGC 3553

4-NGC 3554

5-NGC 3561

6-MGC 5-27-11

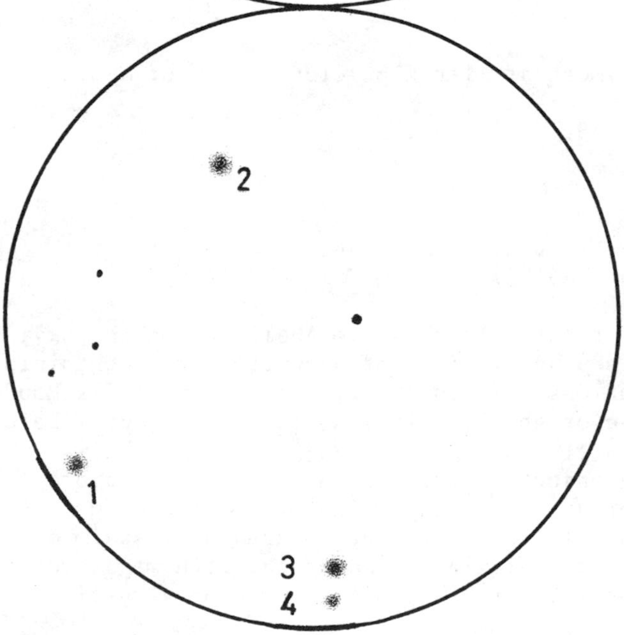

1-NGC 3554

2-NGC 3558

3-MGC 5-27-11

4-NGC 3561

Visual Observations

Abell 1185 is an interesting object visually possessing one field containing six faint galaxies. The following observations were made using a 16" Newtonian reflector.

NGC.3550 Small and irregularly round, about 30" in diameter. Slowly brighter to the middle. Between two stars. Quite bright and brightest of a field of six galaxies.

NGC.3552 Close SP 3553. Almost stellar. Quite bright and much brighter to the middle.

NGC.3553 Similar to, but slightly superior to NGC.3552.

NGC.3554 Small and round. Quite bright and slowly brighter to the middle. Close south of NGC.3552/3553.

NGC.3558 Small and round, about 20" in diameter. Quite bright and slowly brighter to the middle.

MCG 5-27-11 Very small, almost stellar glow close north of NGC.3561.

NGC.3561 Similar to NGC.3558.

7. Abell 1228

Abell	RA	Dec.	AD	AR	Z	RS
1228	$11^h18.9'$	$34^o37'$	1	1	0.034	L

Abell 1228 is situated relatively close to Abell 185 on the sky, has a similar redshift, and is possibly part of a supercluster containing both clusters as condensations. It is a very good example of a Rood-Sastry type L cluster, the brightest galaxies, (all IC objects) being arranged in a well defined line on the sky, about 20' long. The cluster is an interesting object to examine on photographs, since the object MCG 6-25-57, the brightest galaxy, might well be a CD galaxy in an uncatalogued background cluster, being surrounded by a swarm of tiny nebular images. The brightest galaxies shine at the 15th magnitude and are very small, making the cluster difficult to pick up visually.

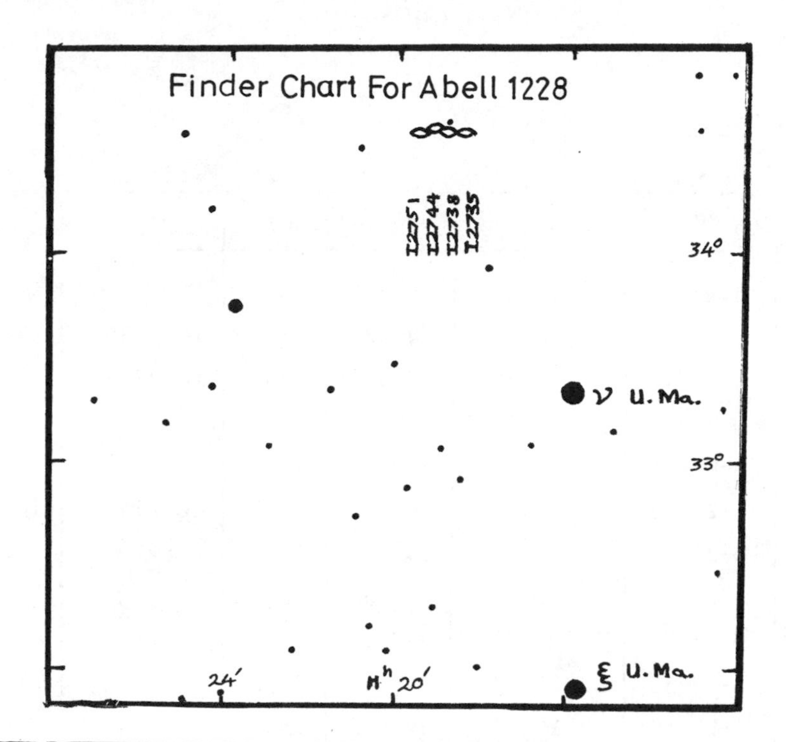
Finder Chart For Abell 1228
I2751
I2744
I2738
I2735
34°
ν U. Ma.
33°
24′
11h 20′
ξ U. Ma.

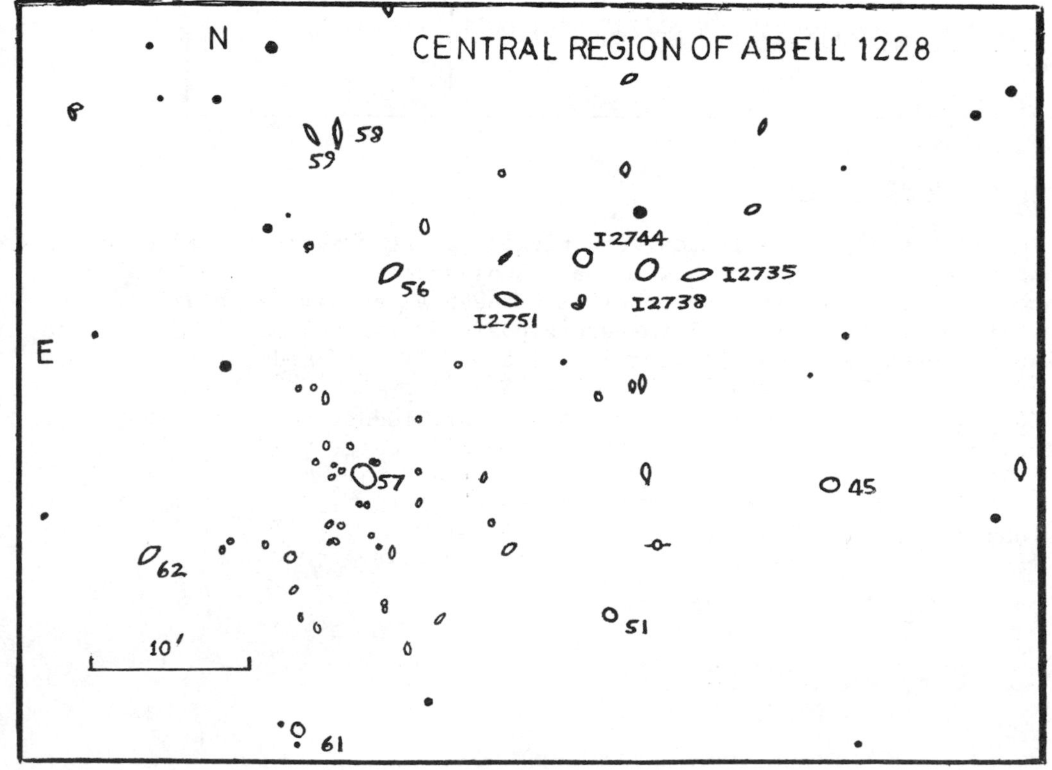
N
CENTRAL REGION OF ABELL 1228
58
59
I2744
I2735
56
I2738
I2751
E
57
45
62
51
10′
61

List of MCG Galaxies

MCG	RA	Dec.	Mp	IC
6-25-45	$11^h17.65'$	$34^o22'$	15	
6-25-48	18.4'	$34^o22'$	15	IC 2735
6-25-49	18.65'	$34^o36'$	15	IC 2738
6-25-50	18.7'	$33^o59'$	15	
6-25-51	18.8'	$34^o11'$	15	
6-25-52	19.0'	$34^o38'$	15	IC 2744
6-25-54	19.4'	$34^o35'$	15	
6-25-56	20.2'	$34^o36'$	15	
6-25-57	20.3'	$34^o22'$	14	
6-25-58	20.4'	$34^o46'$	14.5	
6-25-59	20.5'	$34^o45'$	14.7	
6-25-61	20.75'	$34^o06'$	15	
6-25-62	21.5'	$34^o18'$	15	

Visual Observations

Only two cluster galaxies were picked up after a determined sweep of the area using a 16 inch Newtonian. Subsequent comparison with a large scale 48 inch Schmidt plate confirmed these as anonymous galaxies, F the main cluster, clearly, the wrong position was scanned. The cluster should reveal all the galaxies in the above list in moderate apertures.

MCG 6-25-56 Faint, small and almost stellar, identified from wide double star oriented in PA 45° NF by about 8'. Difficult.

MCG 6-25-57 Very faint and slowly brighter to the middle. Slightly elongated approximately E/W.

8. Abell 1367

WS	Abell	RA	Dec	AD	AR	Z	RS
	1367	$11^h41.9$	$+20^o07$	1	2	0.022	F

Abell 1367[(1), (2)] is one of the richest of the nearby Abell clusters of galaxies, and is a spectacular galaxy hunting ground for medium to large amateur apertures. It contains more than sixty galaxies bright enough to be visually observable in an area of little more than a square degree. Indeed, although A1656, the Coma cluster contains richer galaxy fields, A1367 contains more galaxies brighter than the 14th photographic magnitude. Gregory and Thompson[(2)] have presented strong evidence that A1367 and A1656 may probably be condensations in a vast supercluster: this was arrived at by obtaining redshifts for some of the knots and small clusters of galaxies which lie along a line connecting the two clusters. Some of these knots are visually spectacular and are described in a later part of this volume.

Examination of wide field photographic material shows that the largest and brightest galaxies in the cluster, NGC.3805 (14.0), NCG.3816 (13.5), NGC.3842 (13.3), NGC.3861 (14.0), NGC.3862 (14.0) and NGC.3884 are spread haphazardly across its field. Many subconcentrations are apparent, the largest being around the E galaxy. NGC.3842, which has a spectacular halo of minor galaxies surrounding it. The field around this galaxy is one of the most impressive in the whole sky and contains at least eight visually observable galaxies, including a bright peculiar object, MCG 3-30-66 with photographic magnitude 14.3. This object has recently been shown to be a head-tail radio galaxy by G. Gavazzi[(4)], who also showed that NGC.3842 has a wide-angle tail structure consistent with a possible low orbital velocity with respect to the cluster's centre of mass. It was also shown that the cluster is surrounded by a low brightness diffuse radio source. Abell 1367 is also an X-ray source, and very recent work with the Einstein X-ray observatory has shown that the radiating matter has a clumpy distribution centred around individual knots of galaxies rather than through the cluster as a whole.

The proportion of spirals to elliptical and lenticular galaxies is quite high, as has been pointed out by Dickens and Moss[(1)]. For example, out of the six major galaxies listed above, 3816 is a peculiar spiral (possibly two tidally interacting objects since the 'arms' are counter-wound!), 3861 is possibly an SB, and NGC.3864 shows signs of spirality on a blow-up 48 inch Schmidt wide-field plate.

List of Catalogued Galaxies

The following data, compiled by Malcom Thompson, is based upon the CGCG and cross-correlated with the NGC, NGC, MCG and the UCG.

Chart	RA	Dec.	Mp	NGC/IC	MCG	UCG
1	$11^h38.71'$	$20^o14'$	15.3		3-30-43	
2	39.2'	$20^o23'$	13.6	NGC.3816	3-30-46	0665/50 pec.
3	39.5'	$20^o34'$	13.8	NGC.3821	4-28-30	
4	39.6'	$20^o15'$	15.5			
5	39.6'	$20^o19'$	15.7			
6	39.6'	$20^o22'$	15.6		3-30-48?	
7	39.8'	$20^o24'$	14.7		3-30-51	
8	40.1'	$20^o18'$	15.0		3-30-55	
9	40.3'	$20^o14'$	15.6			
10	40.3'	$20^o21'$	15.4			
11	40.4'	$19^o55'$	15.5		3-30-58	06680/Sb
12	40.6'	$20^o01'$	15.2		3-30-59	06683/SO
13	40.6'	$20^o17'$	15.7			
14	40.8'	$20^o01'$	15.0	IC 2951	3-30-61	06688/Sa
15	40.9'	$19^o54'$	15.2		3-30-62	
16	41.0'	$19^o53'$	15.7		3-30-64	
17	41.1'	$20^o18'$	15.7			
18	41.2'	$20^o15'$	14.3		3-30-66	06697/Irr.

Chart	RA	Dec.	Mp	NGC/IC	MCG	UCG
19	$11^h41.3'$	$20^o10'$	14.2	NGC.3837	3-30-68	
20	41.3'	$20^o03'$	15.2		3-30-69	06901/E
21	41.3'	$20^o13'$	15.3			
22	41.3'	$20^o21'$	14.7	NGC.3840	3-30-70	06702/Sa
23	41.3'	$20^o28'$	15.5			
24	41.4'	$20^o03'$	15.5		3-30-71	
25	41.4'	$20^o05'$	15.7			
26	41.4'	$20^o13'$	13.3	NGC.3842		06704/E
27	41.14'	$20^o15'$	15.0	NGC.3841		
28	41.4'	$20^o18'$	14.9	NGC.3844	3-30-69	06704/SO/a
29	41.5'	$20^o00'$	15.7			
30	41.5'	$20^o16'$	15.1	NGC.3845	3-30-74	
31	41.6'	$20^o07'$	15.3		3-30-76	
32	41.6'	$20^o30'$	15.1			
33	41.7'	$20^o06'$	15.4			
34	41.7'	$20^o15'$	15.2	NGC.3851	3-30-77	
35	41.8'	$20^o00'$	15.5			
36	41.8'	$20^o06'$	15.5		3-30-79	
37	41.8'	$20^o22'$	15.5			
38	41.9'	$20^o20'$	14.9		3-30-83	
39	42.1'	$20^o02'$	15.6			
40	42.1'	$20^o03'$	15.4			
41	42.1'	$20^o09'$	15.5			
42	42.2'	$19^o49'$	15.1	NGC.3857	3-30-84	
43	42.2'	$19^o53'$	15.7			
44	42.2'	$19^o58'$	15.7		3-30-85	
45	42.2'	$20^o05'$	14.5	NGC.3860	3-30-88	06718/Sa/b
46	42.2'	$20^o24'$	14.6		3-30-89	
47	42.3'	$19^o44'$	14.9	NGC.3859	3-30-91	06721/Pec

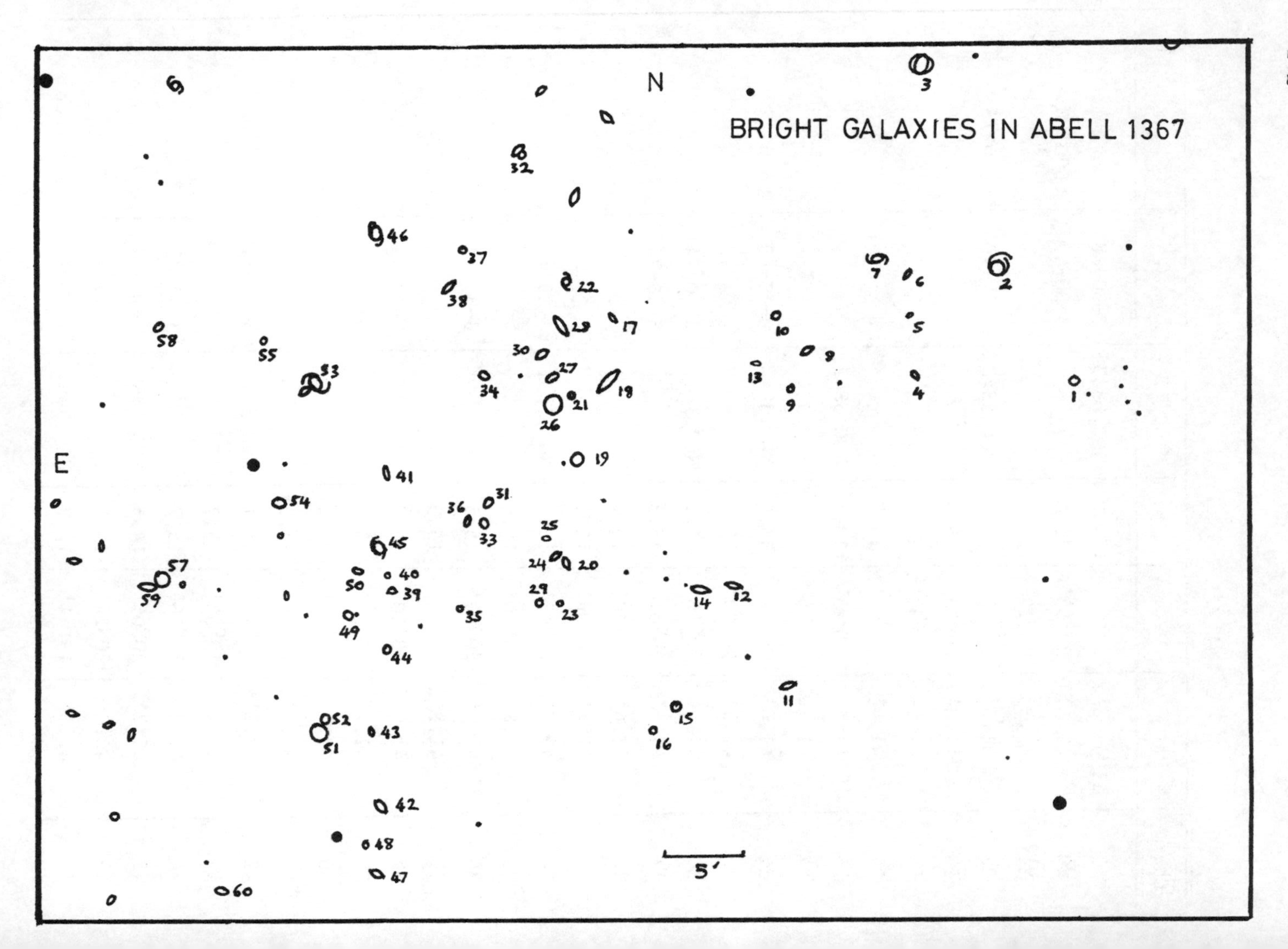
BRIGHT GALAXIES IN ABELL 1367
N
E
5′

Chart	RA	Dec.	Mp	NGC/IC	MCG	MCG
48	$11^h42.3'$	$19^o46'$	15.7			
49	42.3'	$20^o00'$	15.3			
50	42.3'	$20^o03'$	15.6		3-30-92	
51	42.5'	$19^o53'$	14.0	NGC.3862	3-30-95	06723/E
52	42.5'	$19^o54'$	15.2	IC 2955	3-30-96	
53	42.5'	$20^o15'$	14.0	NGC.3861	3-30-93	06724/Sb
54	42.6'	$20^o07'$	15.1		3-30-98	
55	42.7'	$20^o18'$	15.7			
56	42.8'	$19^o43'$	14.8	NGC.3868	3-30-104	
57	43.1'	$20^o03'$	14.2	NGC.3873	3-30-106	06735/E
58	43.1'	$20^o18'$	15.5			
59	43.2'	$20^o02'$	14.8	NGC.3875	3-30-105	06739/SO/a
60	43.3'	$19^o48'$	15.4		3-30-107	
61	43.4'	$19^o43'$	15.6		3-30-108	

Visual Observations

The following visual observations of Abell 1367 were made using the 36" Macdonald Observatory reflector and 16" telescopes situated on both sides of the Atlantic. In all 27 galaxies were observed with the smaller telescopes out of the total of about 60 accessible objects.

MCG 3-30-59
(16) Small and round. Quite bright and brighter to the middle. Close preceeding IC 2951.

IC 2951
(16) Small and slightly elongated E/W. Uniform surface brightness. Prominent star at or near following tip.

Chart (17)
(36) Very small, round almost stellar galaxy a little brighter to the middle.

MCG 3-30-66
(36) Very elongated, conspicuous nebula oriented NP/SF, brightening to the middle.

(16½) Faint but easy to see. Elongated NP/SF being quite narrow and of uniform surface brightness.

(16) Elongated streak about 1' of arc in length. Quite bright and slightly brighter to the middle.

NGC.3837
(36) Small and round, but one of the largest galaxies in the field of NGC.3842.

(16½) South and slightly preceding NGC.3842. Round, faint and brighter to the middle to a stellar nucleus with a small envelope.

(16) Third of 8 and 3rd brightest after NGC.3842 and MCG 3-30-66. Small and round (20"). Quite bright and much brighter to the middle.

Chart 21
(36) Faint almost stellar galaxy between NGC.3842 and MCG 3-30-66.

NGC.3840
(16) Fourth and most northerly of 8, being the northern most of a chain of four small galaxies aligned due north of the fifth member NGC.3842. Small and round. Quite bright and brighter to the middle.

NGC.3842
(36) Brightest object in the central region. Round and bright, being much brighter to the middle.

(16½) Quite bright and irregularly round brightening considerably to the centre to a stellar nucleus surrounded by a diffuse envelope. Brightest galaxy in a field of seven. Faint star close SF.

(16) Brightest and largest of a field of 8. Round (about 45" in diameter). Quite bright and much brighter to the middle. Very faint star close SF.

NGC.3841
(36) Very small and round and much brighter to the middle.

(16½) Almost directly north and close to NGC.3842. Faint and almost stellar.

(16) Close north of NGC.3842. Very small, round, quite bright and brighter to the middle.

NGC.3844
(36) Small and round. Much brighter to the middle.

(16) Second in a chain of 5 from 3840 to 3842. Small and round. Quite bright and brighter to the middle.

NGC.3845
(36) Very similar to NGC.3844.

(16½) North following NGC.3841. Irregularly round and small. Brighter than 3841.

(16) Third on a chain. Similar to the others.

NGC.3851
(36) Very small, almost stellar. Star close preceding.

(16½) North following NGC.3842 is a pair of faint stellar objects. The following of the pair is actually a compact galaxy:- NGC.3851.

(16) Following of a pair of stellar objects.

MCG 3-30-83
(16) Small nebula SP MCG 3-30-89 at a distance. Round, quite bright and brighter to the middle.

NGC.3857
(16) First of a MP field of 4 galaxies. Small (about 20") round quite bright and brighter to the middle. Bright star SF by 3 or 41.

NGC.3860
(16) Small elongated oval of almost uniform low surface brightness in a field containing a few suspected but unconfirmed galaxies.

MCG 3-30-89
(16½) North preceding NGC.3861 at a distance. Faint, quite large and diffuse. Apparent brightening at SP end.

(16) Oval of uniform surface brightness with major axis of about 1' of arc. In field of MCG 3-30-83 which is SP.

NGC.3859
(16) 2nd of a field of four galaxies containing NGC.3862 and first of a field of four galaxies around NGC.3868. Small and round but quite bright and brighter to the middle. Bright star 3 or 4' NF.

NGC.3862
(16) Brightest of a field of 4 being round with a diameter of about 30" and much brighter to the middle.

IC 2955
(16) Very close NP NGC.3862. Small and round with a diameter of about 20". Quite bright and brighter to the middle.

NGC.3861
(16½) Fairly large and bright, being irregularly round and brighter in the middle to a bright nucleus, possibly stellar. Bright star south following.

(16) Quite large, diffuse, of almost uniform surface brightness. Suddenly much brighter to the middle to a tiny stellar nucleus.

MCG 3-30-98
(16) First of a field of 3 galaxies containing NGC.3873. Small and round but quite bright and brighter to the middle. Brighter star NF 3'.

NGC.3868
(16) In a field of four galaxies. Small and round, about 20" in diameter. Quite bright and brighter to the middle.

NGC.3873
(16) Brightest of a field of 3 and of a very close pair (with NGC.3875). Round, having a diameter of about 25". Very much brighter to the middle with a stellar nucleus.

NGC.3875
(16) Small and round, much brighter to the middle to a stellar nucleus. Close SF NGC.3873.

MCG 3-30-107
(16) Small and round, but quite bright and slightly brighter to the middle. N of NGC 3-30-108.

MCG 3-30-108
(16) Oval with major axis of about 30". Uniform surface brightness. Overall an easy object.

NGC.3884
(16) Irregularly round and quite large (about 45" diameter). Quite bright and brighter to the middle to a small core region.

NGC.3886
(16) Round, about 45" in diameter. Quite large and much brighter to the middle.

IC 732
(16) Faint nebula elongated approximately N/S. North preceding NGC.3884 by about 4 or 5' or arc.

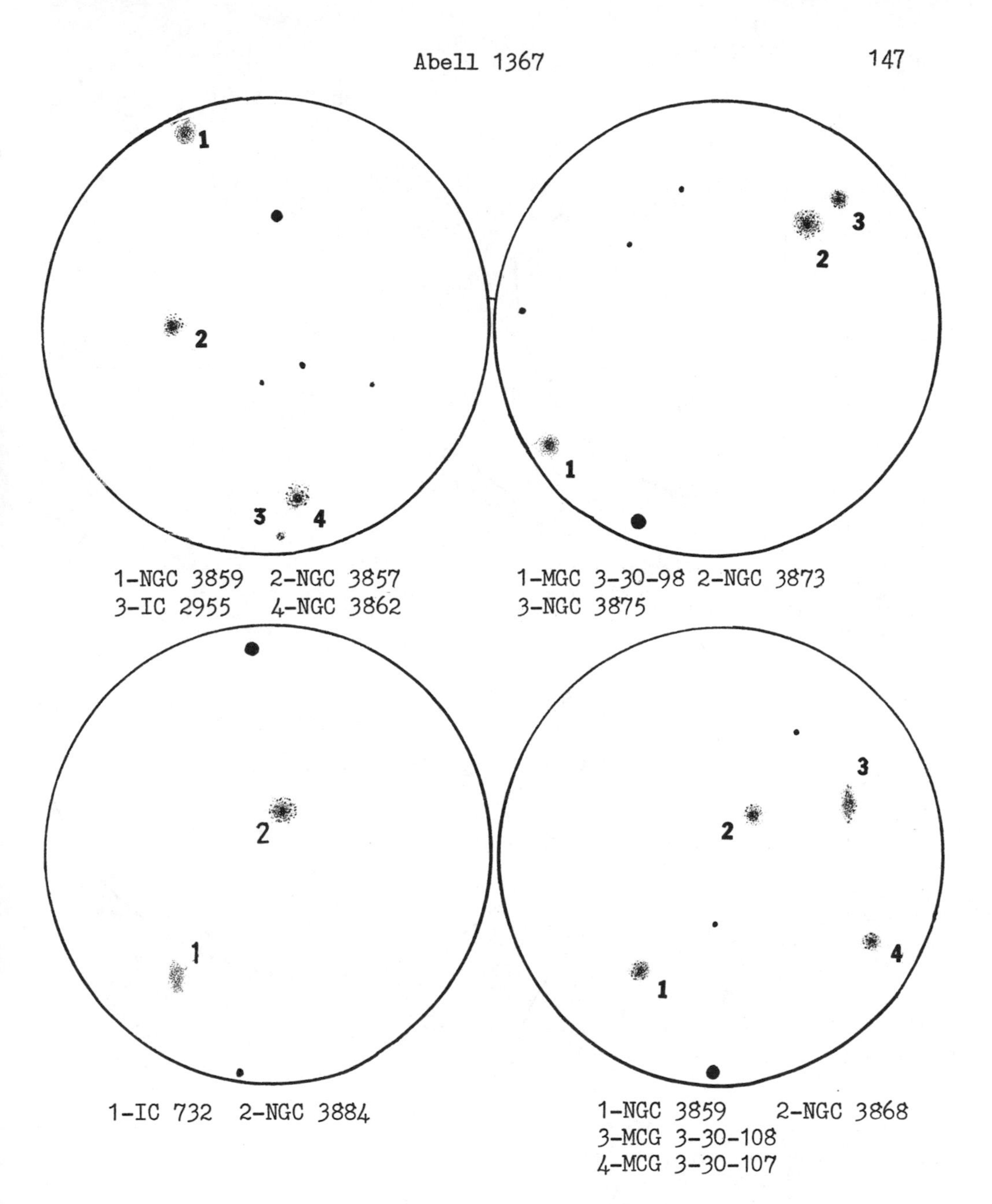

1-NGC 3859 2-NGC 3857
3-IC 2955 4-NGC 3862

1-MGC 3-30-98 2-NGC 3873
3-NGC 3875

1-IC 732 2-NGC 3884

1-NGC 3859 2-NGC 3868
3-MCG 3-30-108
4-MCG 3-30-107

GSW 16-inch f/5 (12')

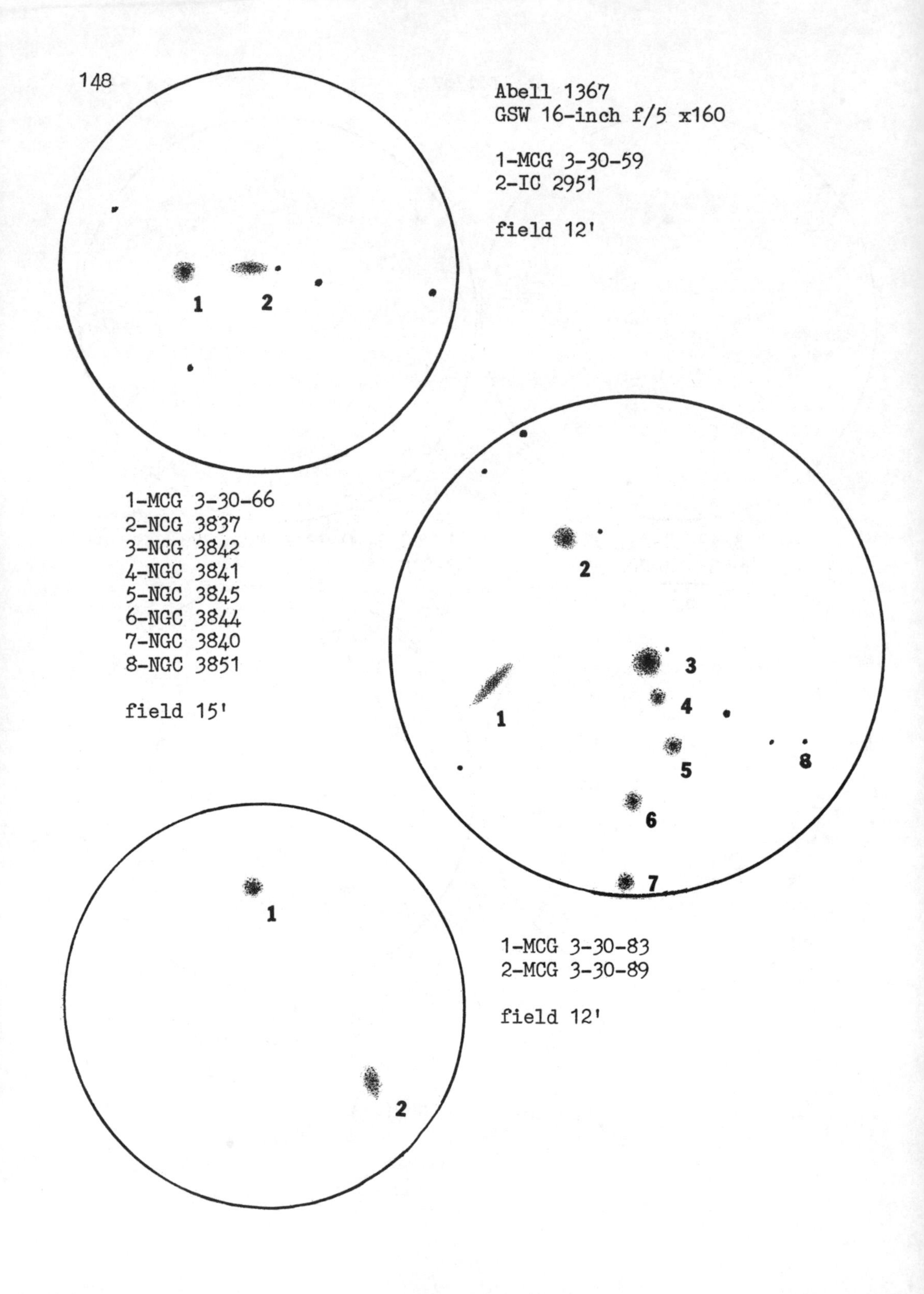
Abell 1367
GSW 16-inch f/5 x160
1-MCG 3-30-59
2-IC 2951
field 12'
1
2
1-MCG 3-30-66
2-NCG 3837
3-NCG 3842
4-NGC 3841
5-NGC 3845
6-NGC 3844
7-NGC 3840
8-NGC 3851
field 15'
1
2
3
4
5
6
7
8
1-MCG 3-30-83
2-MCG 3-30-89
field 12'
1
2

9. Abell 1377

Abell	RA	Dec.	AD	AR	Z	RS
1377	$11^h 44.4$	$+56^o 01'$	3	1	0.0516	C?

Otherwise known as the Ursa Major I cluster, Abell 1377 has a redshift of 0.0516, which makes it the second most distant cluster in this catalogue. Adopting $H_o = 50$ km sec^{-1} Mpc^{-1} implies a distance of 310 Mpc or almost exactly 1,000,000,000 light-years. The cluster was discovered by Baade who measured magnitudes for 42 cluster galaxies. Baade's numbering scheme is followed in the chart of the central region. His photometry has been extended by Godwin and Peach(1), and by Hoffmann and Crane(2), the latter obtaining magnitudes for 305 galaxies including outlying members. The cluster is at the very limit of visibility in a 16 inch f/5 Newtonian reflector, and should be invisible in smaller instruments unless used in exceptionally dark and transparent skies. Moreover, its observation is hampered by the presence of the 6th magnitude star BD + 56 1544 projected near its core, the galaxies being swamped by its brilliant light, scattered in the eyepiece, unless it is kept out of field.

List of Members Brighter Than $V_{25} = 16.0$

Baade	V_{25} Limits
07	14.5-15.0
15	15.5-16.0
20	15.5-16.0
24	14.5-15.0
25	15.0-15.5
26	15.5-16.0
36	"
44	"
48	"
49	"
50	"

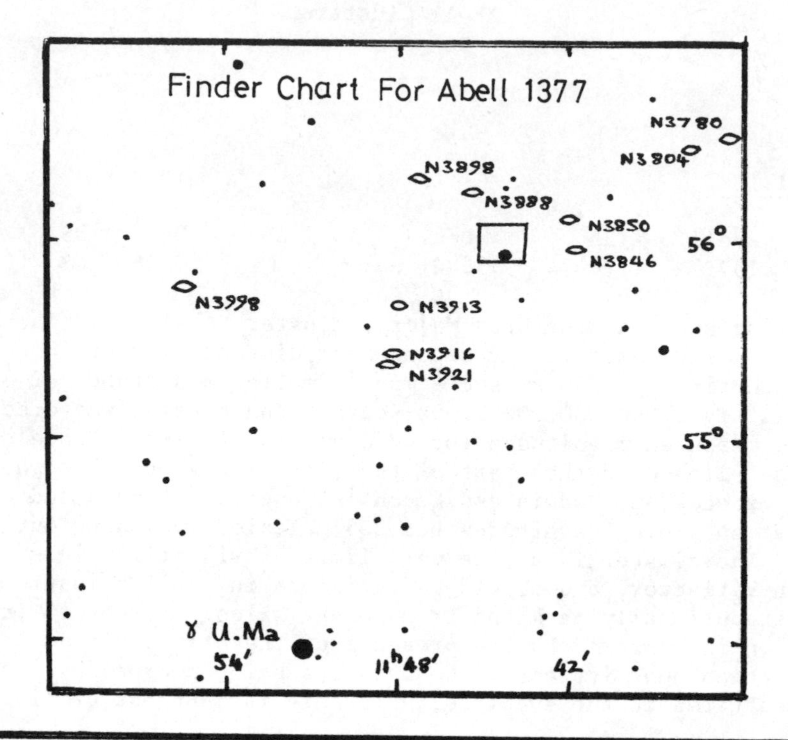

Finder Chart For Abell 1377
N3780
N3804
N3898
N3888
N3850
56°
N3846
N3998
N3913
N3916
N3921
55°
γ U.Ma
54′
11h48′
42′

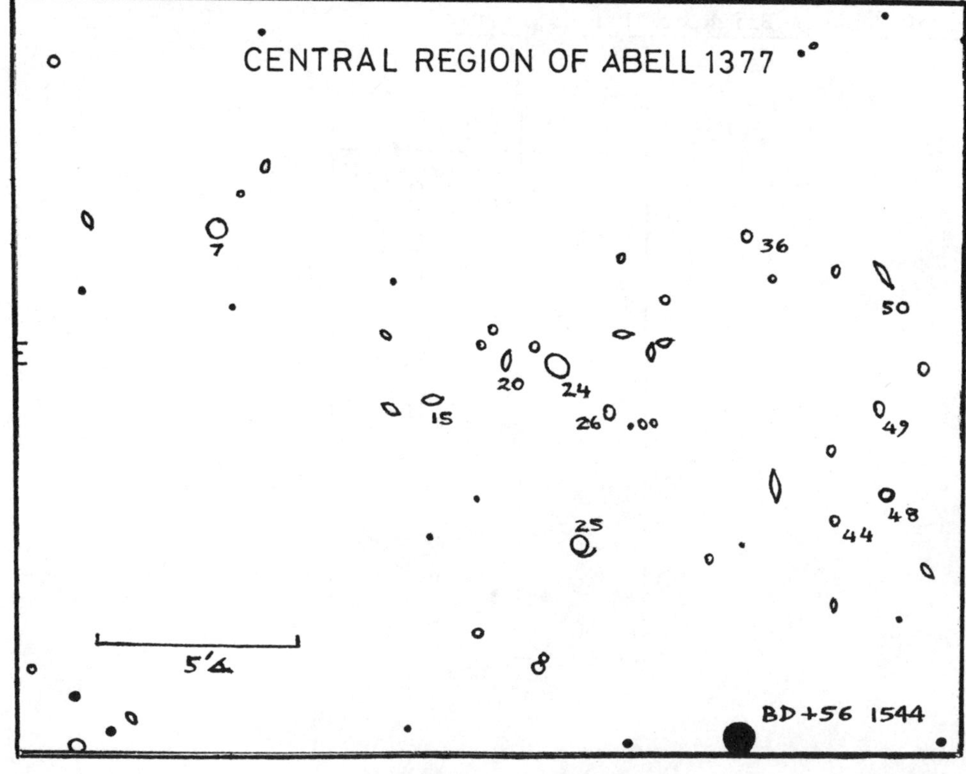

CENTRAL REGION OF ABELL 1377
7
36
50
20
24
15
26
49
48
25
44
5′
BD+56 1544

Abell 1656

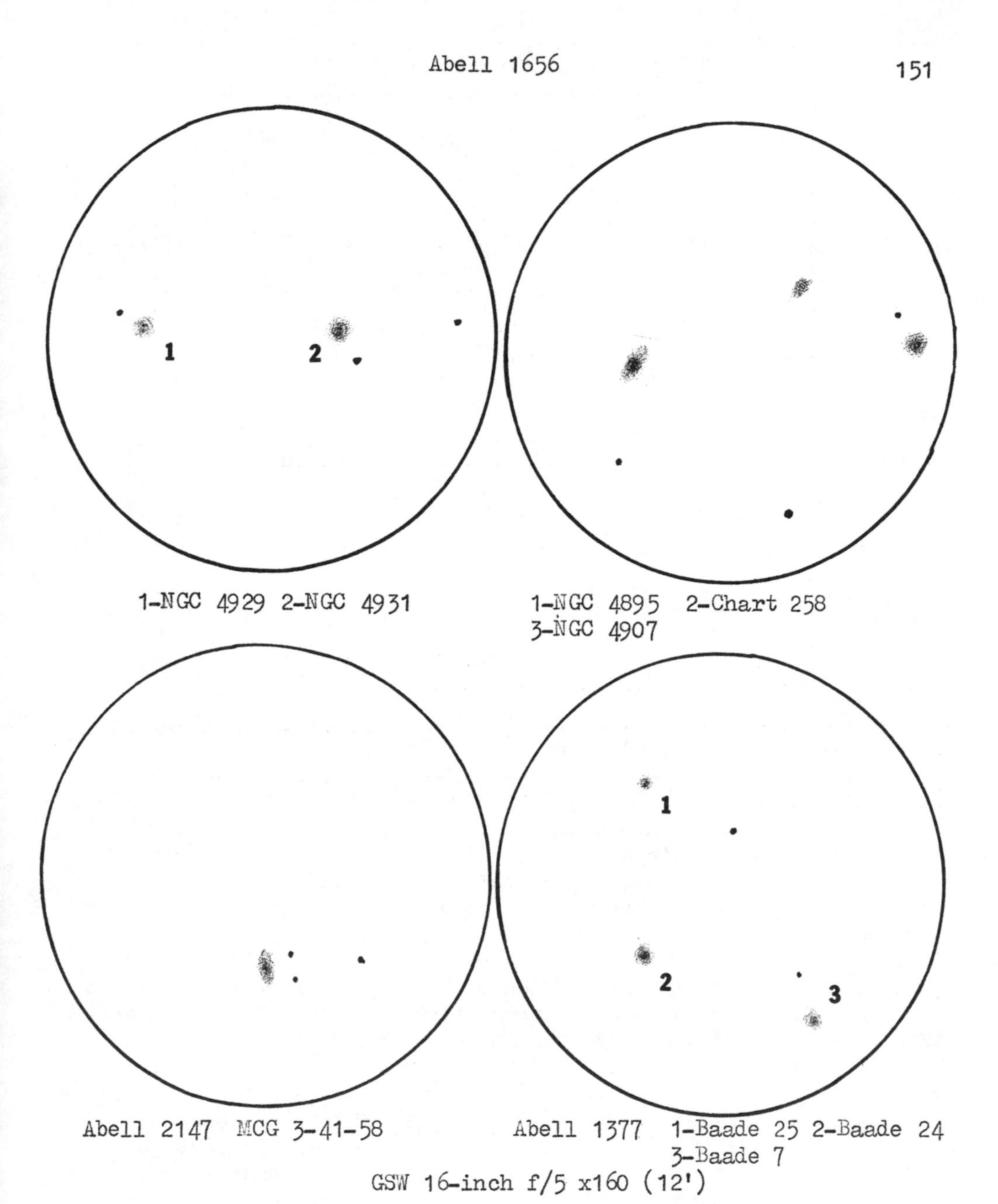

1-NGC 4929 2-NGC 4931

1-NGC 4895 2-Chart 258
3-NGC 4907

Abell 2147 MCG 3-41-58

Abell 1377 1-Baade 25 2-Baade 24
3-Baade 7

GSW 16-inch f/5 x160 (12')

Visual Observations

Baade 07

(16) Very faint, small and irregularly round. Star SP 3'. Difficult.

Baade 24

(16) Similar to Baade 07, but perhaps a little easier.

Baade 25

(16) Extremely faint, almost stellar. Very faint star north following by 2'.

10. Abell 1656

Abell	RA	Dec.	AD	AR	Z	RS
1656	$12^h57.4$	$28^o15'$	1	2	0.023	B

The Coma Bernices Cluster, A1656 is by far the richest and densest galaxy hunting ground for medium to large amateur apertures. The best comparison of its richness is with A426 or A1367 for which observations of respectively 18 and 27 cluster galaxies appear in this volume. In the Perseus cluster, the most crowded field, that around NGC.1275, contains nine galaxies. The field of NGC.3842 in A1367 contains eight galaxies, other fields containing three or four galaxies. However, in A1656, the field of NGC.4874 contains eighteen galaxies, the field of NGC.4889 contains thirteen, and several other fields contain six or seven galaxies. In all, observations of about fifty central galaxies will be described below.

A1656 contains a high proportion of E and SO galaxies to spirals in its core, although not as high as proportion as A426. Its brightest galaxies are NGC.4874 and 4889, the latter being slightly brighter than the former. This pair of galaxies dwarfs all other cluster members in size and leads to its Rood-Sastry classification as a binary 'B' cluster. The EO galaxy NGC.4874 is surrounded by a dense halo of compact E and SO galaxies, and NGC.4889 has a similar, if less populous halo. The centre of mass of the cluster is usually taken to lie midway between the above giant galaxies, and the cluster is roughly spherically symmetrical about this point. Godwin et al.[1] have recently constructed the luminosity function for the cluster, obtaining photometric apparent visual magnitudes for about 1000 galaxies brighter than the 19th magnitude in an area of about 3.25 square degrees. The cluster has been traced to about 6° on angular distance - redshift plots by Chincarini and Rood[2], whilst Gregory and Thompson[3] have presented evidence that A1656 and A1367, some 11° away in the sky, may

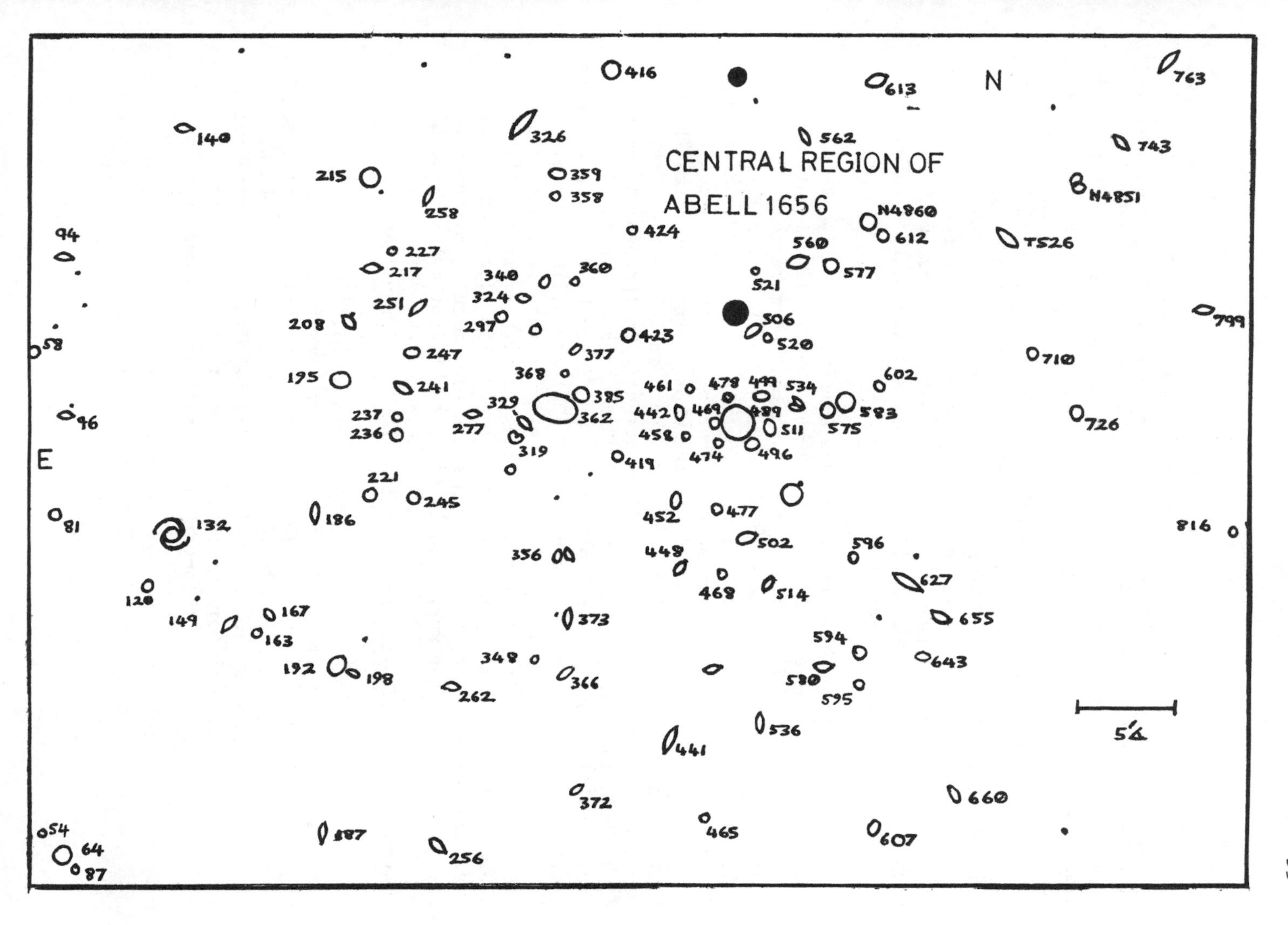
CENTRAL REGION OF
ABELL 1656
N
E
5'

be part of a supercluster - this work being corroborated by Chincarini and Rood[4] using Sc galaxies as distance indicators. The distribution of galaxies in the Coma cluster is shown in figure (8), based upon a chart in Godwin et al.[1]. Figure (9) shows the average radial distribution obtained by ring counts on figure (8) compared with a King model. The central density of galaxies is about 2,500 galaxies per square degree, corresponding to a space density of about 1000 galaxies per cubic megaparsec, using the redshift distance of 130 Mpc for H_o = 50 km sec^{-1} Mpc^{-1}. The characteristic core radius obtained from the King Model is R_c = 0.38 Mpc. Thompson[5] has detected a degree of systematic radial alignment of galaxies in the core of the Coma cluster, although this is a marginal effect, not at all apparent on casual inspection of plate material.

Many observations have indicated the presence of an intracluster medium in this cluster. A1656 is an X-ray source, almost half as luminous as A426. It is also a radio source, and more than 20 individual galaxies have been detected using the Westerbork radio telescope[6]. The most interesting are associated with NGC.4874 and NGC.4869. The former galaxy shows a double structure at a scale of 15" whilst the latter is a head-tail source whose morphology gives the impression of the orbit of the galaxy around the cluster centre. Another way that intracluster matter has been detected is through the detection of a 'cold spot' in the universal 2.7° K microwave background radiation[7]. This marginal effect was predicted by Soviet workers to occur in situations where the microwave photons collide with electrons in the intracluster matter, and lose energy to the electrons, hence being shifted to slightly longer wavelengths with a slightly lower characteristic blackbody temperature. Abell 1656 also contains excellent examples of Van den Bergh's anaemic spirals, stripped of interstellar gas, the most notable being the bright galaxy NGC.4921.

List of Central Galaxies Brighter than the 16th Magnitude

The following data is taken from Godwin et al.[1] and is a list of all the galaxies on the chart of the central region with integrated visual magnitude greater than 16. Note that, in[1], integrated magnitudes are those integrated out to the 25 mag. arc sec^{-2} isophote; these may be taken as the total integrated magnitude for our purposes. The identification numbers given on the chart are also due to Godwin et al., and are cross-correlated with the NGC and IC only.

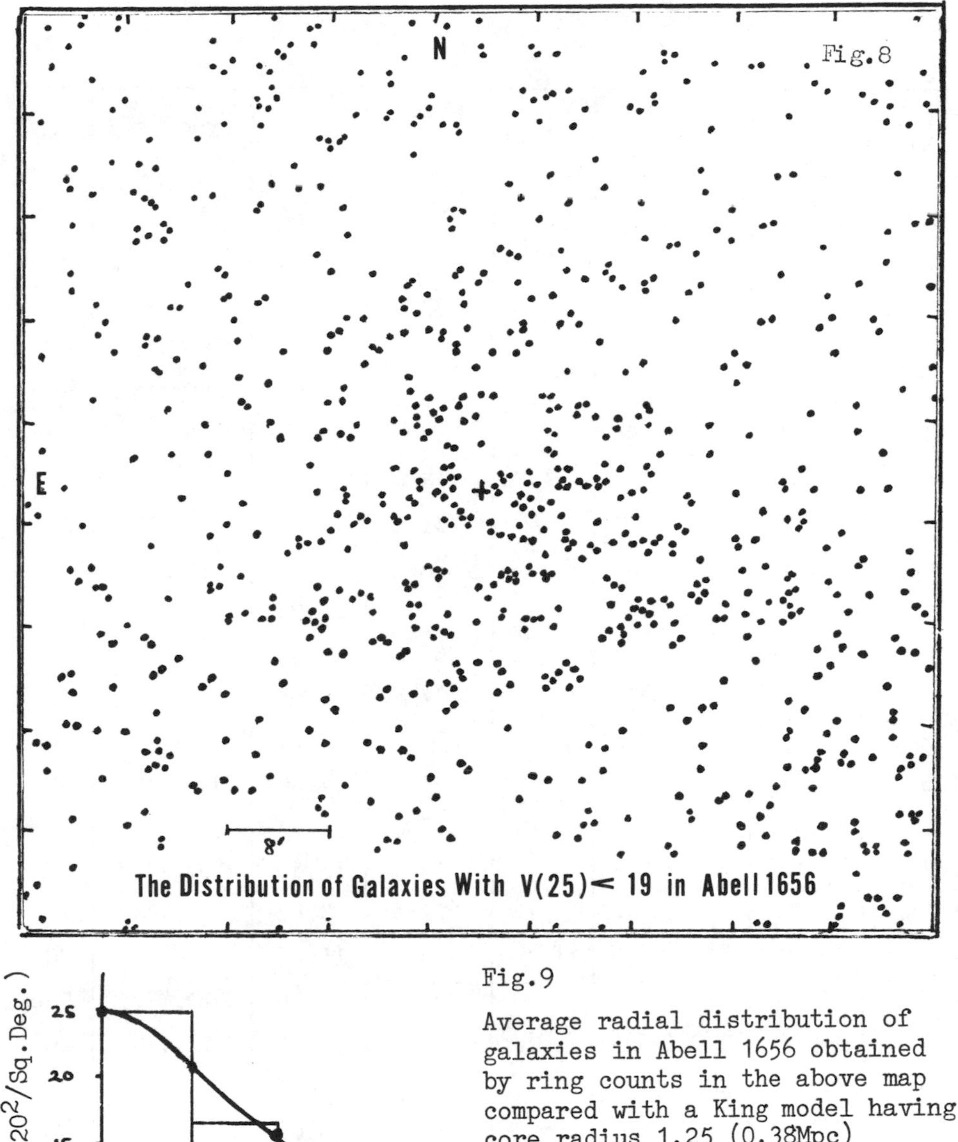

Fig.8

The Distribution of Galaxies With V(25) < 19 in Abell 1656

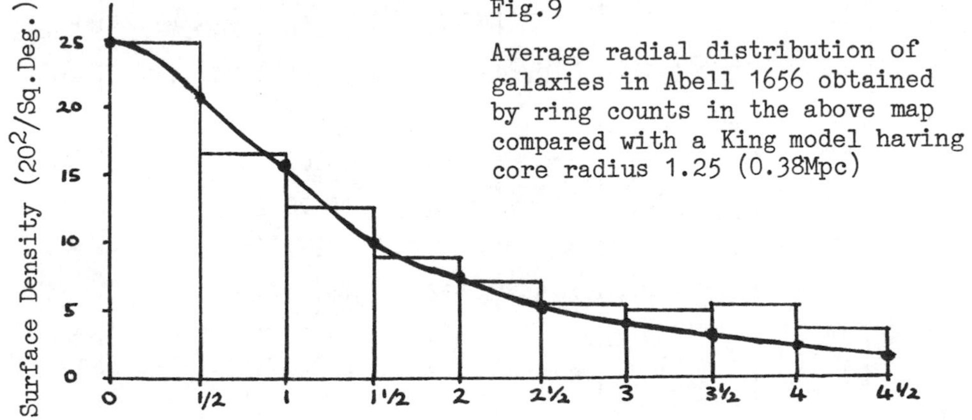

Fig.9

Average radial distribution of galaxies in Abell 1656 obtained by ring counts in the above map compared with a King model having core radius 1.25 (0.38Mpc)

Galaxy Number	V_{25}	NGC/IC	Type	Galaxy Number	V_{25}	NGC/IC	Type
54	15.49			234	14.04		
58	13.68	NGC.4927		236	14.99		SO
64	13.05	NGC.4926	E	237	14.21	IC 4042	SO
81	14.47			241	14.75	IC 4041	E
87	15.53			245	14.29	NGC.4906	E
94	14.28			247	15.36		SO
96	15.67			248	15.62		
120	13.70	NGC.4923		251	14.83	IC 4040	I
130	15.70		E	256	15.31		
132	12.46	NGC.4921	S	258	14.98		
140	14.97			262	15.70		
149	13.84	NGC.4919	SO	277	15.07		SO
152	15.64		E	279	15.21		SO
156	15.54			297	14.60	IC 4026	SO
163	15.10			313	15.83		SO
167	15.08		E	319	15.50	NGC.3898	E
171	15.25			322	13.82	NGC.4989	E
186	14.04		SO	324	15.42		SO
187	15.44			326	13.20	NGC.4895	Pec
192	13.05	NGC.4911	S	329	15.06	NGC.4894	SO
195	13.39	IC 4051	E	332	14.80	IC 4021	E
198	15.57			340	15.02		SO
208	13.78	NGC.4908	E	348	15.73		
209	13.88			356	15.51		S
215	13.57	NGC.4907	S	358	15.89		SO
217	14.04	IC 4045	E	359	14.93		E
221	15.41		SO	360	14.75	IC 4012	E
227	15.53		SO	362	11.78	NGC.4889	E

Galaxy Number	V_{25}	NGC/IC	Type	Galaxy Number	V_{25}	NGC/IC	Type
366	15.02			506	14.33		SO
368	15.24	IC 4011	E	511	14.19	NGC.4871	SO
372	15.50			514	14.65	IC 3976	E
373	14.94			520	15.07		E
377	15.03		SO	521	15.91		SO
385	13.91	NGC.4886	E	533	15.76		SO
416	13.60	NGC.4881	E	534	15.49		E
419	15.28		SO	536	15.22		SO
423	14.27	NGC.4883	SO	538	15.80		SO
424	15.30		SO	544	13.85	NGC.4869	E
441	14.03			550	15.80		
442	14.78	IC 3998	SO	551	15.47		SO
448	14.93		SO	558	15.38		E
452	14.46	NGC.4876	E	560	13.60	NGC.4865	E
458	15.81		SO	562	15.74		SO
461	15.21		E	575	14.43	NGC.4867	E
464	15.53		E	577	14.56		SO
465	15.55			580	14.66	IC 3963	SO
467	15.38		E	583	14.19	NGC.4864	E
468	15.53		SO	587	15.41		E
469	15.87		SO	594	14.16	IC 3959	E
474	15.54		SO	595	14.79	IC 2957	E
477	14.78	NGC.4875	SO	596	14.66	IC 3960	SO
478	15.73		SO	602	14.43	IC 3955	SO
489	12.20	NGC.4874	E	607	14.67		S
496	14.29	NGC.4872	SO	609	15.43		SO
499	14.42	NGC.4873	SO	612	15.07	NGC.4858	S
502	14.32	IC 3973	SO	613	14.23		

Galaxy Number	V_{25}	NGC/IC	Type	Galaxy Number	V_{25}	NGC/IC	Type
619	15.90		SO	709	15.66		E
623	15.08			710	14.03		E
624	15.94			722	15.55		
627	14.81	IC 3949	Pec	726	14.20	NGC.4850	SO
631	15.45		SO	727	14.85		
632	15.54			739	14.86		
637	15.46			743	14.88		
640	15.81		SO	763	13.73	NGC.4848	Pec
643	14.74	IC 3947	SO	765	15.80		
650	15.29		SO	768	15.67		
655	14.15	IC 3946	SO	770	14.93		
660	14.00	NGC.4854	SO	779	15.24		
690	13.54	NGC.4853	E	816	15.34		
702	15.68		E	818	14.54		

Comments on Identifications

NGC.4851 This identification is taken from Rood and Baum[7], the galaxy thus identified is a double system and does not appear in the list of Godwin and Peach. The RNGC magnitude is Mp = 15.0, (2E).

NGC.4860 Godwin and Peach identify NGC.4858 as the southern of a double galaxy, typing it as a spiral. They do not list the northern component designated NGC.4860, Mp = 14.5 with Dreyer description F of DNEB.

T 526 This designation was taken from a paper by Strom and Strom[8], the galaxy not being mentioned in Godwin and Peach.

RNGC 4882 According to Sulentic and Tifft, NGC.4882 is a 15.5 mag. object situated 3' south of NGC.3883 and 0.2' time preceding NGC.4886. The only object visible on a 16" × 20" blow-up of a 48" Schmidt plate of the central region of A1656 near this position is near mag. 19. Hence RNGC 4882 is a misidentification and should be deleted from the catalogue.

Visual Observations

The following sub-catalogue of visual observations of the Coma Cluster was made with telescopes ranging from 8½ inch to 36 inch and is meant to cover only the central region of the cluster as depicted upon the main chart. One or two objects lying just outside this field are also included.

NGC.4927
(16) Quite a bright elongated nebula, oriented approximately N/S. Nucleus offset to north. Small and brighter to the middle. (Galaxy + star.)

NGC.4926
(16) Quite bright and slowly brighter to the middle. Slightly elongated. About 1' × 30". Anon galaxy NP, GP 44, at V_{25} = 14.53. Small and irregularly round. Preceding this galaxy is GP54, faint and almost stellar.

NGC.4923
(16) South following NGC.4921. Small (less than 30"), round and quite bright, brightening slowly to the middle.

NGC.4921
(16½) Bright central core in large diffuse envelope.

(16) Brightest of a field of 4 galaxies. Relatively large, comparable to 4889 and 4874 in size and brightness, being just over 1' in diameter. Round with a diffuse halo brightening slowly to the middle.

(8) Very difficult. Just visible at high power.

NGC.4919
(16) SP NGC.4921 and 4923 in a field of 4. Similar to 4923 but slightly inferior in size and brightness.

Chart 171
(16) North following of two in a medium power field. Round and small (less than 30"). Slightly superior to GP 234 preceding in the same field.

NGC.4911
(16) Round with a diameter of about 30". Quite bright and brighter to the middle. Star of mag. 13 close preceding and another, much fainter, close following. RNG 4911A = GP 198 attached to SP edge of 4911 not seen.

(8½) Faint large blur requiring averted vision. No condensation. Difficult.

IC 4051
(16) Small and round, quite bright and brighter to the middle. Eastern of a field of seven galaxies south of NGC.4907. About 14.5 mag.

NGC.4908
(16) Similar to IC 4051

NGC.4907
(16) Small and round, approximately 20" in diameter. Brightest of a field of seven galaxies. A star of mag. 13 (RNGC) situated close SP.

IC 4045
(16) Faint and almost stellar. About mag. 15.

Chart 234
(16) Small preceding GP 171 in a MP field. Small and round (25"). Quite bright and brighter to the middle.

IC 4042
(16) Small and round (20"). Quite bright and brighter to the middle. Superior to IC 4041 situated close NP. About mag. 15.

Chart 258
(16) Small and slightly elongated NS. Faint.

NGC.4906
(16) An almost stellar haze surrounding the following component of a faint double star.

IC 4040
(16) Similar to IC 4045 in the same MP field.

Chart 277
(36) Small and round conspicuous nebula with a stellar galaxy close south.

(Chart 297
(36) Small, and faint nebula. An anonymous almost stellar galaxy preceding. Forms a group with very faint almost stellar galaxies 324, 340 and 360.

NGC.4898
(36) Very close double galaxy. A pair of almost stellar nebulae with the following the brighter having a stellar nucleus.

(16) An easy object south following NGC.4889 in a field of 13 galaxies. Round and small (less than 30"). Quite bright and brighter to the middle.

NGC.4895

(16½) Easily visible in a field of many galaxies. Irregularly round, perhaps elongated NP/SF. RNG. 4895A lies SP, an almost stellar nebulous knot. (RNGC 4895A = Chart 359?)

(16) Quite bright and brighter to the middle. Slightly elongated with a PA of about 30°. About 1' × 30". One of the easiest objects.

NGC.4894

(36) Small and round, brightening to a stellar nucleus.

(16½) Very faint, almost stellar haze south following NGC.4889.

(16) Faint and almost stellar nebula south following NGC.4889.

NGC.4889

(36) Relatively large and bright oval nebula elongated E/W and brightening considerably to a prominent stellar nucleus. Small edge-on galaxy oriented NP/SF almost in contact near preceding tip.

(16½) Bright with a very bright centre. Slightly elongated NP/SF.

(16) Large and bright in comparison with other galaxies in A1656. Brightest nebula in the cluster, just brighter than NGC.4874. Oval, about 1.5' × 1', oriented roughly E/W. Much brighter to the middle.

(10) Elongated almost E/W. Considerably brighter to the middle.

(8) Best seen at MP. Slightly brighter to the middle. Faint stars near.

NGC.4886

(36) Small and round, quite bright to a faint stellar nucleus.

(16) Easy once spotted. Small and round (20"). Quite bright and brighter to the middle. Close NP NGC.4889. Slightly inferior to NGC.4883 NP.

Chart 368

(36) Very small and round. Stellar nucleus. Close NF NGC.4886. Chart 340 lies close north and is a very faint almost stellar object.

NGC.4883

(36) Considerably faint, small round nebula brighter to the middle.

(16) Similar to NGC.4886 but slightly superior. One of 13 galaxies in the field of NGC.4889 and one of 17 in the field of NGC.4874.

Chart 356
(36) Very close double galaxy, both components being almost stellar. The NP component is very faint whilst the SF is extremely faint.

NGC.4881
(16) One of 3 galaxies in the field of NGC.4895. Quite bright and brighter to the middle. About 40" diameter. Very easy.

Chart 419
(36) Small and round galaxy. Extremely faint in line (P) with two stars S of NGC.4889.

(16) Extremely faint object at the limit of vision. Almost stellar.

IC 3998
(36) Small and round, slightly brighter to the middle. Considerably faint.

(16) Galaxy preceding NGC.4874. Faint and almost stellar. Relatively difficult.

Chart 461
(36) Small and round considerably faint. Extremely faint almost stellar galaxy SF.

(16) Almost stellar galaxy NF NGC.4874. Very faint.

NGC.4876
(36) Small and round. Considerably faint but brighter to the middle.

(16) An easy object in the field of NGC.4874. Small and round, quite bright and brighter to the middle.

Chart 474
(36) Southernmost of a line of 3 galaxies 474/469/478. Close SF NGC.4874. All three galaxies being almost stellar and very faint.

NGC.4875
(36) Considerably faint and similar to NGC.4876 preceding.

(16) Faint and almost stellar galaxy in the field of NGC.4874.

<u>NGC.4874</u>
(36) Relatively pretty large, round nebula being considerably bright and brightening considerably to the middle. In a 12' field of 30 galaxies.

(16) Large and bright compared to most other galaxies in A1656. Round (about 1.2'-1.5' in diameter) with a diffuse halo brightening slowly to the middle. Centred in a medium power field of 17 galaxies.

(10) Exceptionally faint, elusive, roughly elliptical haze seen with AV.

(8½) Difficult. Faint extended haze at medium power. More compact than NGC.4889. Fair image at high power.

<u>NGC.4873</u>
(36) Small and round, considerably faint.

(16) Easy once located. Small and round (20"). Quite bright and brighter to the middle.

<u>NGC.4872</u>
(36) Small and round, considerably faint, close SP NGC.4874.

(16) Almost stellar and very close SP NGC.4874. Easy once spotted.

<u>IC 3973</u>
(36) Small and round. Considerably faint.

(16) Faint, almost stellar galaxy SP. NGC.4875. Relatively difficult.

<u>NGC.4871</u>
(36) Small and round. Considerably faint, much brighter to the middle.

(16) Close preceding NGC.4874. An easy object. Small and round (about 20" in diameter). Quite bright and brighter to the middle.

<u>Chart 534</u>
(36) Very close pair of faint, almost stellar nebulae. Almost stellar anon. galaxies NF and NP.

<u>IC 3976</u>
(36) Small and round. Considerably faint.

NGC.4869
(36) Round nebula, slightly larger than average. Considerably faint but much brighter to the middle. Star close NP. Two almost stellar anonymous galaxies SP.

(16½) An easy object SP NGC.4874.

(16) An easy object similar to NGC.4871 but larger and brighter. Star in contact of about mag. 12 or 13.

NGC.4867
(36) A considerably faint, small round galaxy south preceding NGC.4864.

(16) Following of a close double galaxy with NGC.4864. Both rather similar. Small and round, about 25". An easy object being quite bright and brighter to the middle.

IC.3963
(16) Eastern of a group of seven galaxies around IC 3946. Faint and almost stellar but brighter than IC 3957 close preceding.

NGC.4863
(16) Small, round nebula NP a star of magnitude 6 in the central region, quite bright and brighter to the middle.

NGC.4864
(36) A round nebula, relatively quite large, pretty faint and much brighter to the middle. Close NP NGC.4867.

(16) Preceding of a close double galaxy with NGC.4867. Similar to 4867.

IC 3959
(16) 5th of seven galaxies around IC 3946 and second of a compact subgroup of three of which it is the brightest. Very small, about 20" in diameter, round and slowly brighter to the middle. Third brightest in field and about mag. 15.

IC 3957
(16) Fourth of seven and first of a subgroup of three with IC 3959 and IC 3963. Faint and almost stellar, mag. about 15.5.

IC 3955
(36) Considerably faint, small, round nebula.

(16) Very small galaxy, close preceding NGC.4864/4867. Quite bright and brighter to the middle.

NGC.4858
(16) Close SP NGC.4863. Smaller and slightly fainter. Round, quite bright and brighter to the middle.

IC 3949
(16) Third of seven and brightest and largest in the field. About mag. 14.5. Easy to direct vision and slowly brighter to the middle. Long and thin (about 1' × 15"), oriented roughly E/W.

IC 3947
(16) Very faint, about mag. 15.5. Almost stellar. Second of seven.

IC 3946
(16) Faint, very small and only just non-stellar. Round and brighter to the middle. Star of mag. 12 NP by about 1'. First of seven and second brightest in the IC 3949 group.

NGC.4848
(16½) Very faint, small, irregularly round. Slightly brighter centre. At X176 appears extended NP/SF.

(16) Quite a bright spindle about 1.2' × 0.4' oriented roughly N/S. Bright star NF.

NGC.4841
(16½) X84 shows what appears as two nuclei. X176 slightly elongated NP/SF. Also at this power, the second nucleus in the F end shows as an attached star. (This is probably RNGC 4841B = GP 854).

(16) A pair of galaxies in contact, a larger brighter galaxy S of a smaller, almost stellar galaxy = RNGC 4841B. Sketch checks well with 48 inch plate.

NGC.4851
(16) Preceding of a field of 8 galaxies preceding the field of NGC.4874. Small and round, but quite bright and brighter to the middle.

NGC.4854
(16) Quite bright, small and round, about 20". Slowly brighter to the middle. Object possibly a galaxy SP by 6 or 7'. Very small with a relatively bright stellar nucleus. Position corresponds with NGC.4853.

NGC.4929
(16) Small and almost stellar. Quite bright. Star preceding by 1'. RNGC quotes the magnitude of the star as 16.

NGC.4931
(16) Small (about 25") and round. Quite bright and brighter to the middle. Star of magnitude about 13 NF by 1'.

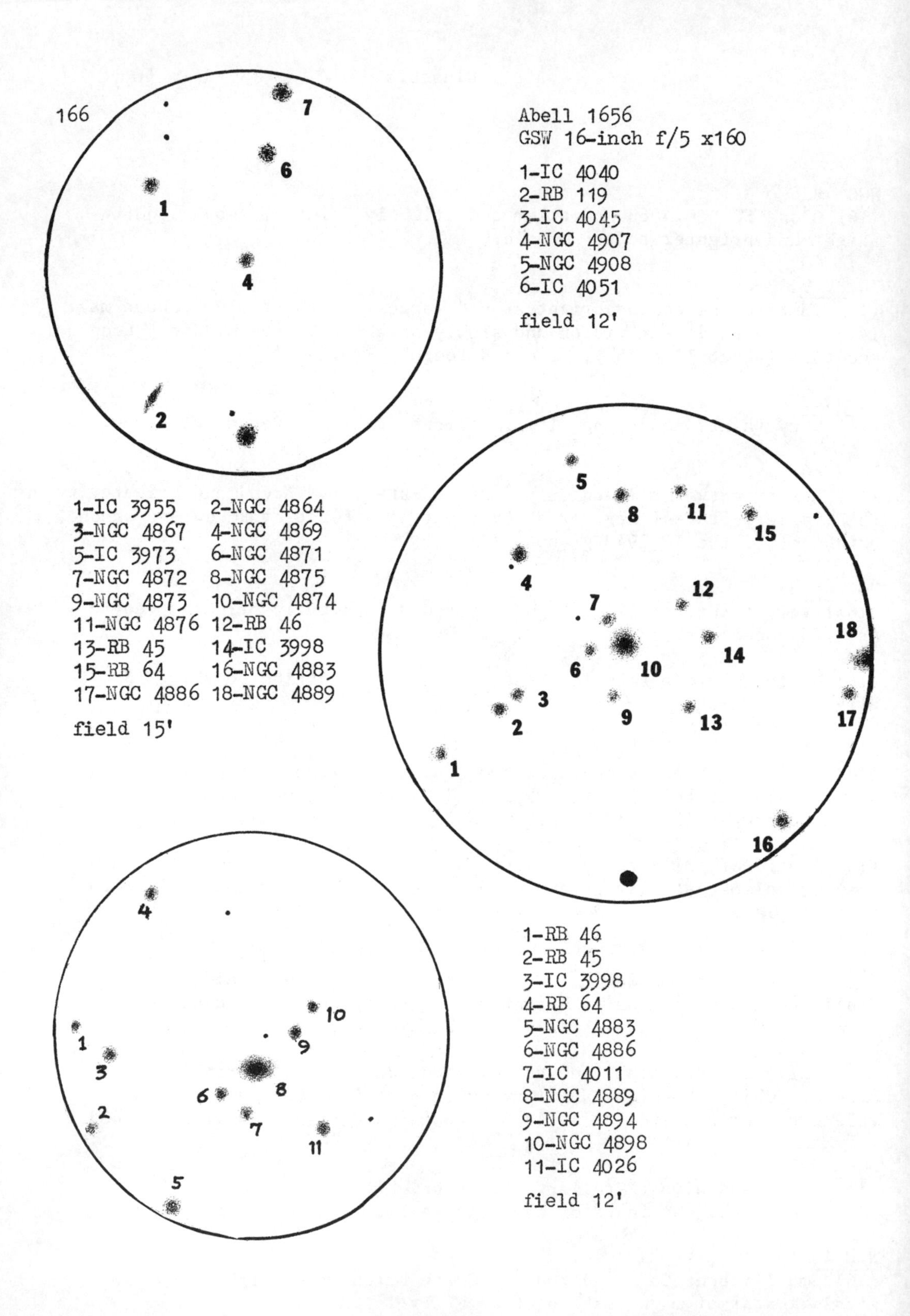
Abell 1656
GSW 16-inch f/5 x160
1-IC 4040
2-RB 119
3-IC 4045
4-NGC 4907
5-NGC 4908
6-IC 4051
field 12'
1-IC 3955 2-NGC 4864
3-NGC 4867 4-NGC 4869
5-IC 3973 6-NGC 4871
7-NGC 4872 8-NGC 4875
9-NGC 4873 10-NGC 4874
11-NGC 4876 12-RB 46
13-RB 45 14-IC 3998
15-RB 64 16-NGC 4883
17-NGC 4886 18-NGC 4889
field 15'
1-RB 46
2-RB 45
3-IC 3998
4-RB 64
5-NGC 4883
6-NGC 4886
7-IC 4011
8-NGC 4889
9-NGC 4894
10-NGC 4898
11-IC 4026
field 12'

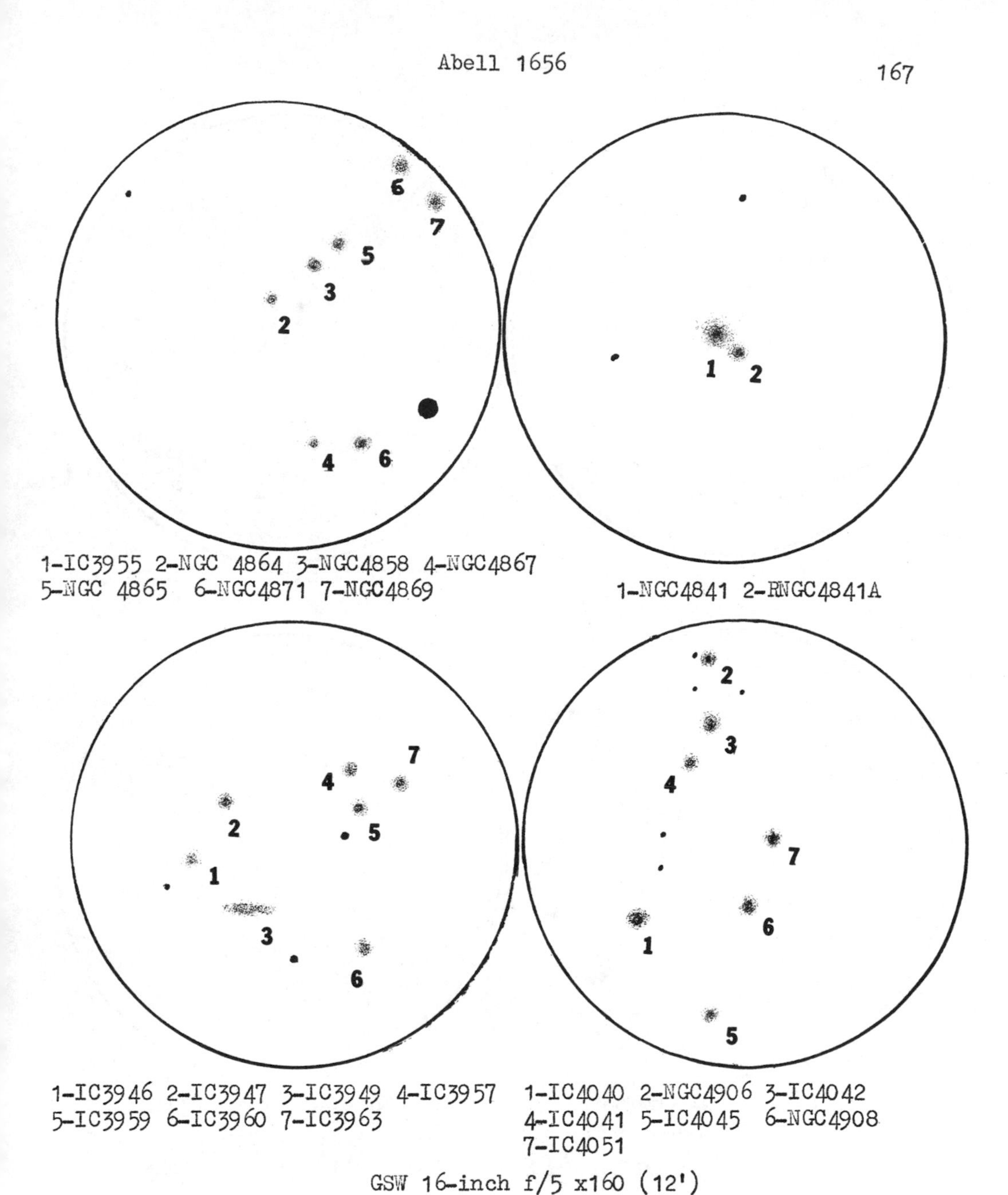

1-IC3955 2-NGC 4864 3-NGC4858 4-NGC4867
5-NGC 4865 6-NGC4871 7-NGC4869

1-NGC4841 2-RNGC4841A

1-IC3946 2-IC3947 3-IC3949 4-IC3957
5-IC3959 6-IC3960 7-IC3963

1-IC4040 2-NGC4906 3-IC4042
4-IC4041 5-IC4045 6-NGC4908
7-IC4051

GSW 16-inch f/5 x160 (12')

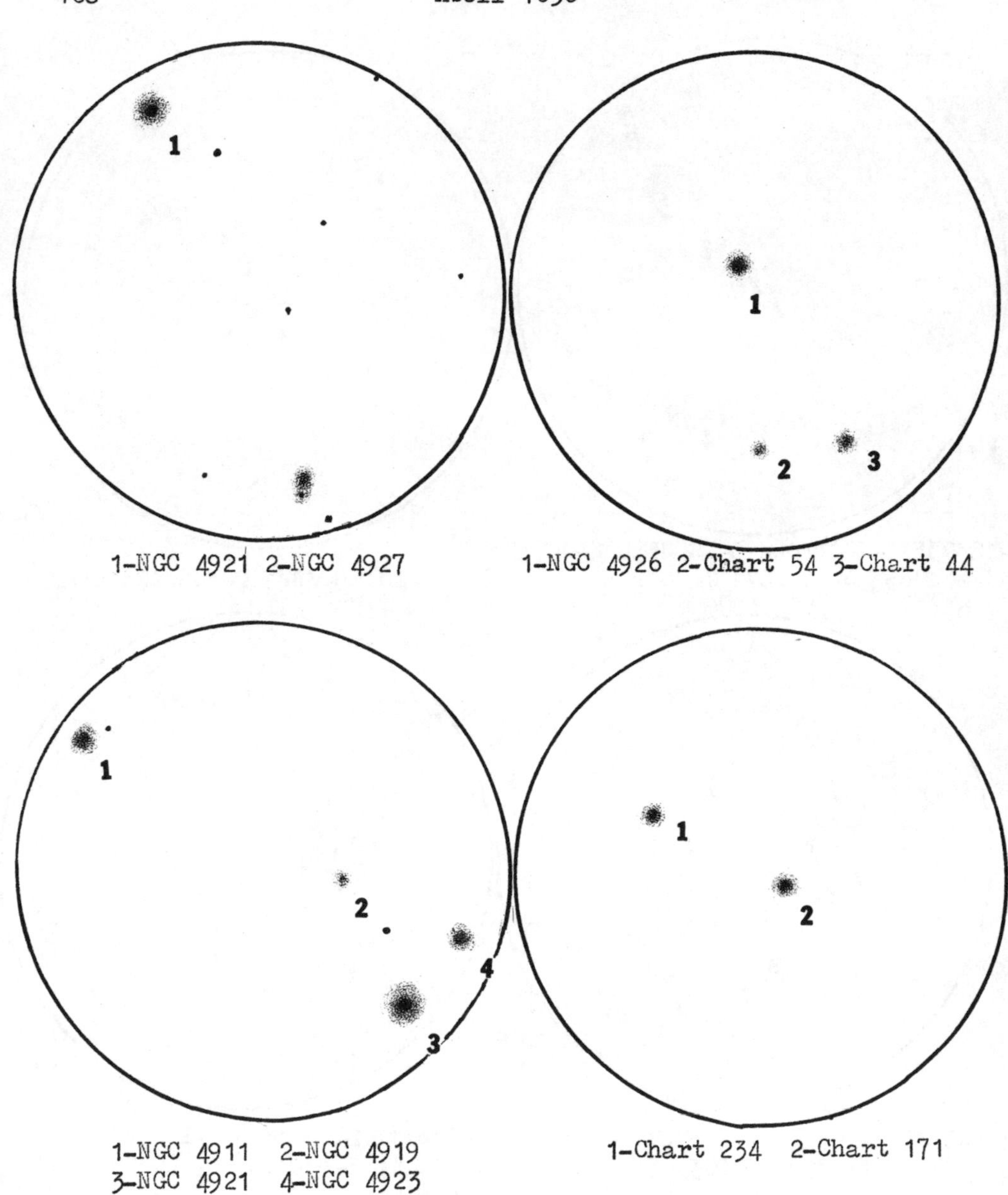

1-NGC 4921 2-NGC 4927

1-NGC 4926 2-Chart 54 3-Chart 44

1-NGC 4911 2-NGC 4919
3-NGC 4921 4-NGC 4923

1-Chart 234 2-Chart 171

GSW 16-inch f/5 x160 (12')

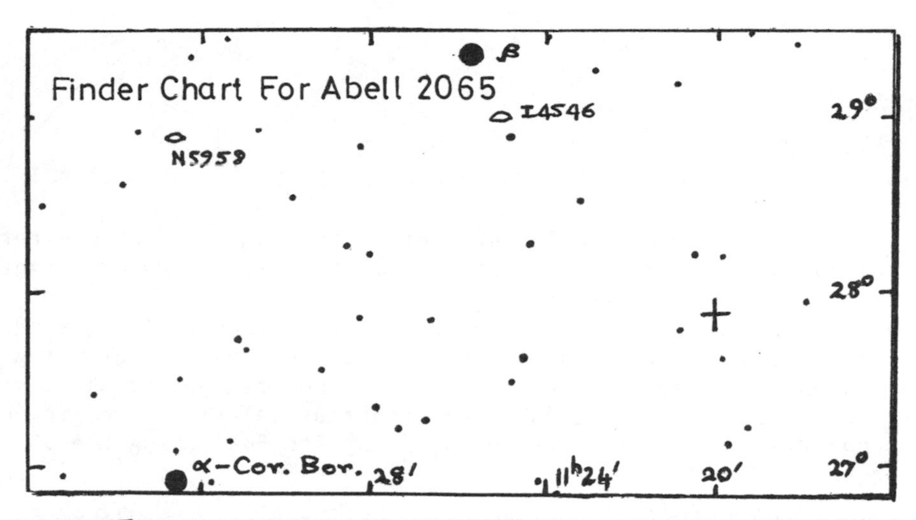
Finder Chart For Abell 2065
β
I4546
N5958
29°
28°
27°
α-Cor. Bor.
28′
11h24′
20′

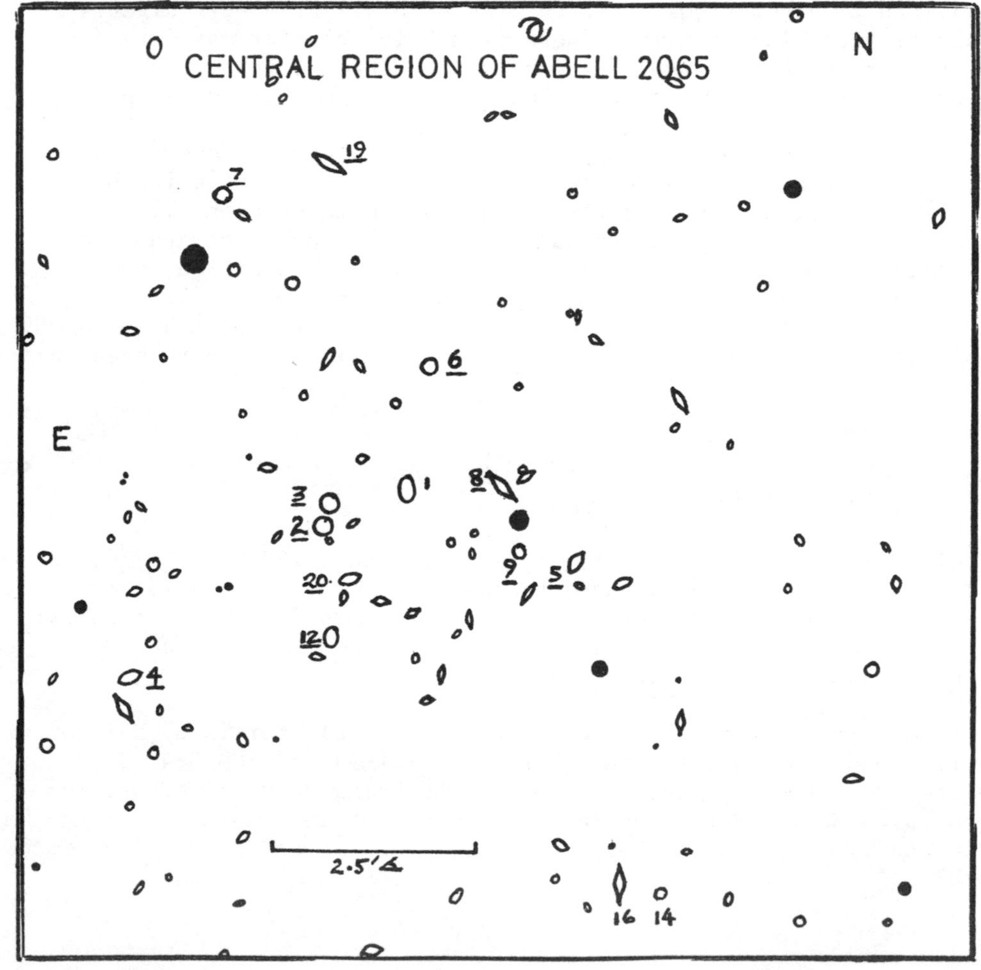
CENTRAL REGION OF ABELL 2065
N
E
19
7
6
3
2
1
8
9
5
20
12
4
2.5′
16
14

11. Abell 2065

Abell	RA	Dec.	AD	AR	Z	RS
2065	$15^h20.0$	+27°50'	3	2	0.0722	C

Abell 2065, the Corona Borealis cluster is the most distant cluster in this catalogue having a redshift distance of about 1,500,000,000 light years. Because of its staggering distance it is beyond the reach of most amateur astronomers, being undetectable in 16 inch reflectors even under good sky conditions. A visual observation of the cluster using a 24 inch Newtonian reflector in extremely dark and transparent skies has, however, been reported[1]. Ten of the brightest galaxies were picked up. The observation reported here was made by Ron Buta using the 36 inch reflector of the University of Texas:

> '36 inch telescope 360X. Sketch made on a slightly hazy night. Only sixteen cluster members were seen. The cluster was very difficult in the 36 inch. Except for the three brightest members, the cluster was a challenge for the 36 inch telescope.'

None of the cluster galaxies appears in any catalogue. However, several studies have been made of the cluster, the most recent of which being that of Godwin et al.[2] whose numbering system is used on the chart of the central region which was prepared from a 200 inch photograph and shows galaxies brighter than V_{25} = 16 to 16.5. Galaxies 1-6 have magnitudes between 15.5 and 16.0, the remaining numbered galaxies magnitudes 16 to 16.5 and all others are fainter. Photographs of A2065 are plentiful in the literature, one of the best is that reproduced in[13].

12. Abell 2147

Abell	RA	Dec.	AD	AR	Z	RS
2147	$16^h00.0$	+16°03'	1	1	0.0377	I

Abell 2147 lies at a redshift distance of about 225 Mpc or 750,000,000 light-years and is situated close to the Hercules cluster A2151 on the sky. The latter is slightly closer, with redshift 0.0360, implying a radial separation of only 10 Mpc. The two clusters form the major sub-condensations of the Hercules Supercluster.

The most interesting object contained in the cluster is a chain of five galaxies, whose brightest members are MCG 3-41-54 and MCG 3-41-51, enveloped in a common halo, the whole chain being about 4' of arc in length (0.26 Mpc). Photographic enlargements of the chain show it to be surrounded by many faint galaxies, possibly background objects but more probably cluster dwarfs.

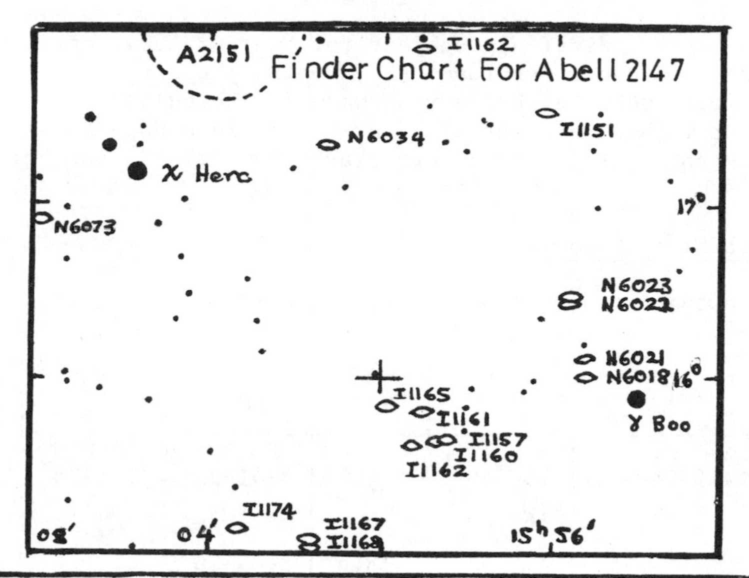
Finder Chart For Abell 2147
A2151
I1162
N6034
I1151
χ Herc
N6073
17°
N6023
N6022
N6021
N6018
16°
γ Boo
I1165
I1161
I1157
I1160
I1162
I1174
I1167
I1169
08'
04'
15h 56'

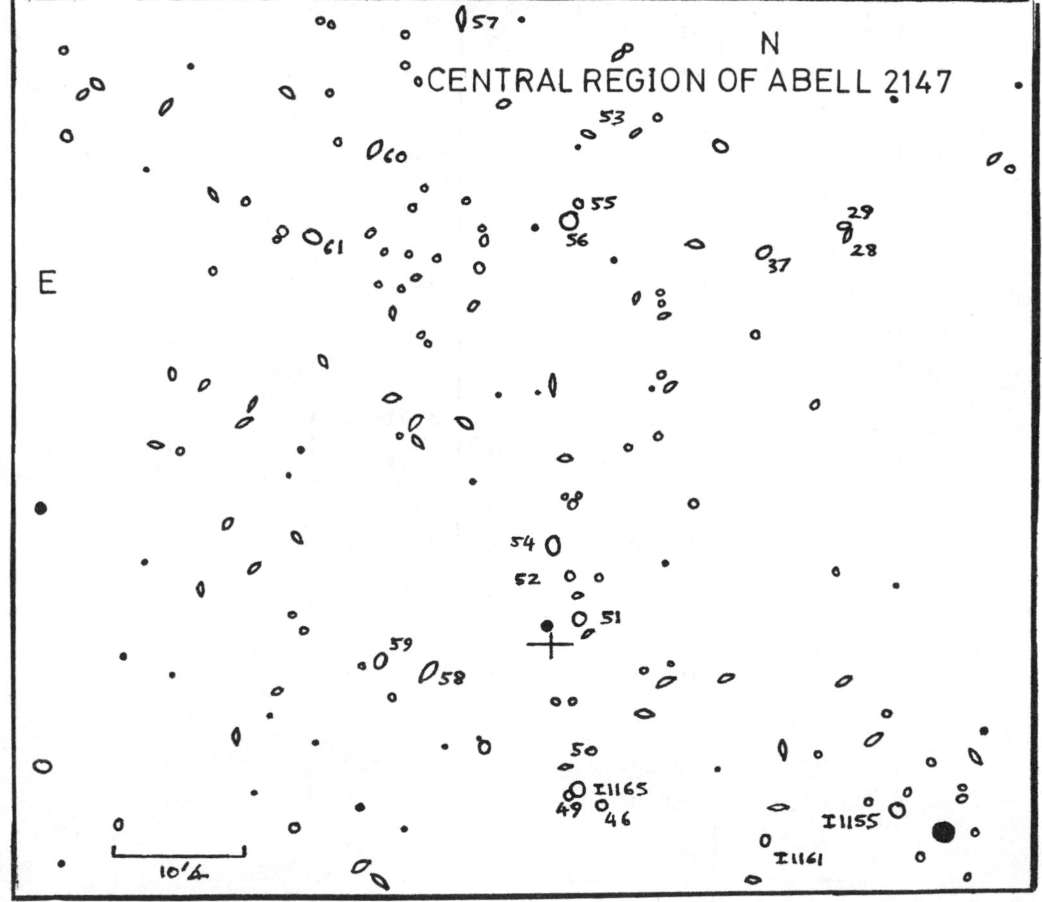
CENTRAL REGION OF ABELL 2147
N
E
57
53
60
55
56
61
29
28
37
54
52
51
59
58
50
I1165
49
46
I1155
I1161
10'4

The cluster is a difficult subject for amateur telescopes, but is easy to find, the above mentioned chain being projected behind a sixth or seventh magnitude star halfway between chi-Herculis and gamma-Bootis. Its brightest members are scattered over its field however and thus the cluster does not contain any spectacular galaxy fields for amateur apertures.

List of Catalogued Galaxies

The following data was compiled from the MCG.

RA	Dec.	Mp	MCG	IC
15^h58.3	15 49	14.9	3-41-23	IC 1155
58.4	16 49	16	3-41-24	
58.45	16 51	15.4	3-41-26	
58.6	16 28	15.0	3-41-28	
58.6	16 29		3-41-29	
59.0	15 47	15.2	3-41-36	IC 1161
59.05	16 26	15.2	3-41-37	
59.1	16 49	15.2	3-41-38	
59.8	15 49	16	3-41-46	
59.9	15 50	14.6	3-41-48	IC 1165
59.9	15 50		3-41-49	
59.9	15 51	17	3-41-50	
59.95	16 03	15.5	3-41-51	
59.95	16 05	15.3	3-41-52	
16^h00.0	15. 34	15.4	3-41-53	
00.05	16 07	14.9	3-41-54	
00.05	16 30	14.6	3-41-55	
00.1	16 29	15.7	3-41-56	
00.6	16 42	15.4	3-41-57	
00.7	15 58	15.1	3-41-58	
00.9	15 59	16	3-41-59	
00.95	16 32	15.5	3-41-60	
01.25	16 27	14.8	3-41-61	

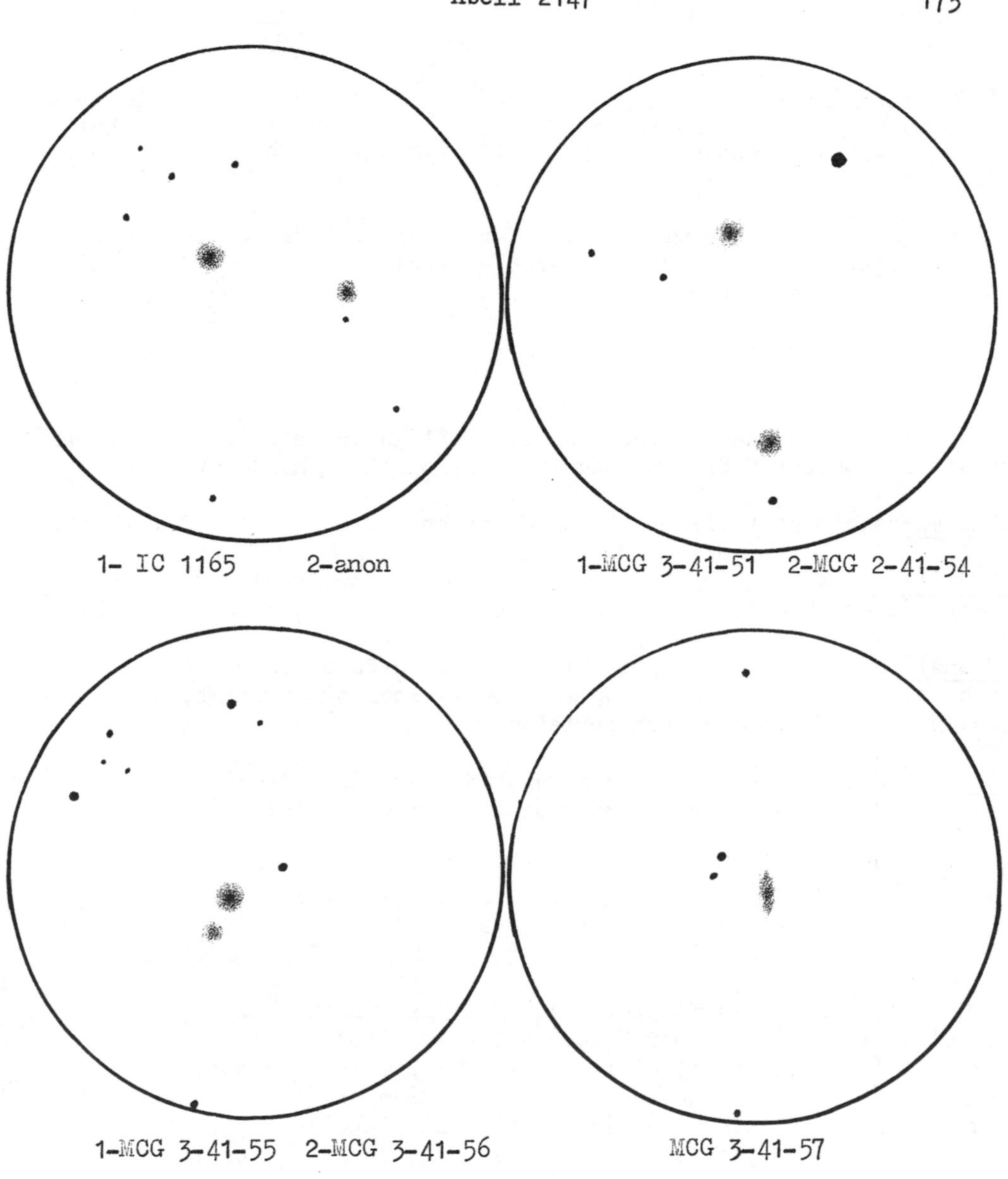

GSW 16-inch f/5 x160 (12')

Visual Observations

The following galaxies were observed with a 16 inch f/5 Newtonian reflector.

IC 1165 Small and round with a diameter of about 25". Quite bright and slowly brighter to the middle. Faint uncatalogued galaxy NF by about 4' with star close north.

MCG 3-41-51 First of two. Small and round, quite bright and brighter to the middle.

MCG 3-41-54 Second and northern of two. Slightly inferior to 3-41-51 but similar object. Star of magnitude 11 or 12 1' N.

MCG 3-41-55 Almost stellar but quite bright. Close N of MCG 3-41-56.

MCG 3-41-56 Small and round, having a diameter of about 20". Quite bright and brighter to the middle. Star of magnitude about 9 following.

MCG 3-41-57 Small and slightly elongated with PA of about -45°. Quite bright and brighter to the middle. Double star preceding by about 3', minor component being extremely faint.

MCG 3-41-58 Small and slightly elongated approximately NS. Quite bright and brighter to the middle. Pair of stars mag. 12 following.

13. Abell 2151

Abell	RA	Dec.	AD	AR	Z	RS
2151	$16^{h}03'$	$+17^{o}50'$	1	2	0.036	I

The famous Hercules Cluster Abell 2151 is a rich, amorphous, spiral-rich cluster similar in morphology and galaxy content to the Virgo cluster. Its Rood-Sastry classification - Irregular - suggests an irregular sprawl, with no central condensation around a central galaxy or galaxies. A glance at a wide field POSS plate(1) substantiates this. Deeper, smaller 5 m plates(1) show a beautiful disarray of intertwined spirals, ellipticals, edge-on systems, and a fascinating object with a jet, IC 1182. The latter is a peculiar galaxy, possibly spiral, from which emerges a chain of knots about 1.5' of arc in length. These may be faint background galaxies accidentally aligned, but the distorted nature of IC 1182 indicates that the jet might be real. Another very interesting object is the pair of interacting galaxies IC 1178 and IC 1181, which on a 48 inch Schmidt plate show huge plumes of material ejected in opposed spiral directions. The whole system is about 3' in diameter implying a true diameter of about 190 Kpc at the redshift distance of 216 Mpc of A2151. NGC.6050 and IC 1179 consists of an apparently closely intertwined pair of Sc spiral galaxies, reminiscent of NGC.4567/8 in the Virgo Cluster.

Abell 2151 contains a few radio galaxies. For example, the galaxies NGC.6034 and NGC.6061 are associated with wide-angle head-tail sources, so the cluster probably contains remnants of the primeval gas from which it condensed. The gas must be distributed in a clumpy fashion because the proportion of spiral galaxies is rather high (60% S galaxies) and is not detected as a diffuse X-ray source(2). In the ram-pressure model, cluster spirals should be stripped of gas by orbital motion through the intracluster medium. However, it is possible that the spiral population inhabits the outermost regions of the cluster preferentially, as in the case of the similar cluster A262.

The visual observation of the cluster is rendered difficult by the high proportion of low contrast type objects, and by its relatively large distance. The easiest objects are NGC.6040 and its close companion NGC.6041, the former shining at Mp 14.5. A moderate aperture (16 inch +) and grim determination together with a familiarity with the field should yield at least a dozen galaxies.

The cluster has been the subject of a detailed visual survey by Ron Buta using the 30 inch and 36 inch reflecting telescopes at Macdonald Observatory of the University of Texas who made the following general remarks.

'The Hercules cluster covers many fields of the 30-inch. At least 50 galaxies in the central band and north-east region could be detected without a finding chart. Other bright members are visible south of the main band. Three members (NGC.6040, 6041 and IC 1181) appeared as conspicuous doubles, clearly resolved into both components. NGC.6050 is also a close double, but in this case the components are both late-type spirals and the fainter component (an Scd) was difficult to see in both the 30-inch and 36-inch telescopes. I only saw both components clearly in an observation made with the 82 inch telescope at a magnification of about 600. The visually brightest galaxy in the central $1.4^{o} \times 1.2^{o}$ of the cluster is the close pair NGC.6041, whose combined visual magnitude must be near 13.0. There are at least 20 other galaxies in this region whose visual magnitudes are near 14 ± 0.5'.

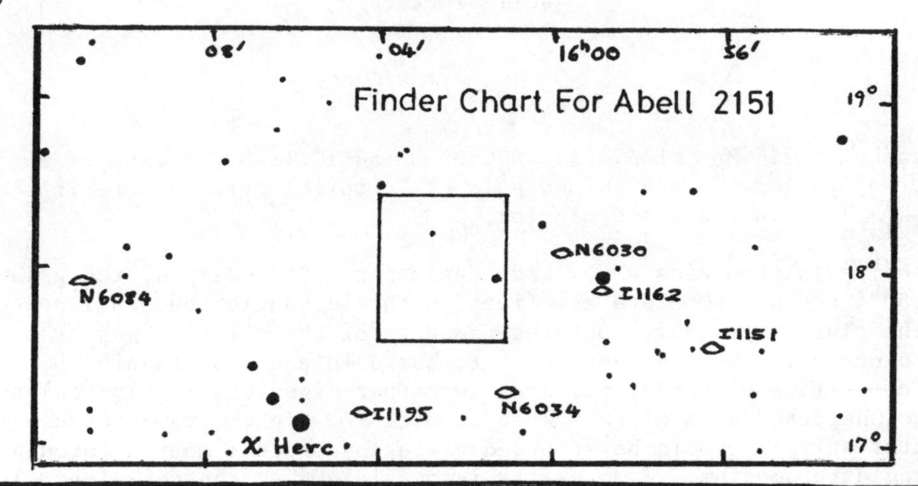

08'
04'
16h00
56'
Finder Chart For Abell 2151
19°
18°
17°
N6030
I1162
N6084
I1151
I1195
N6034
χ Herc

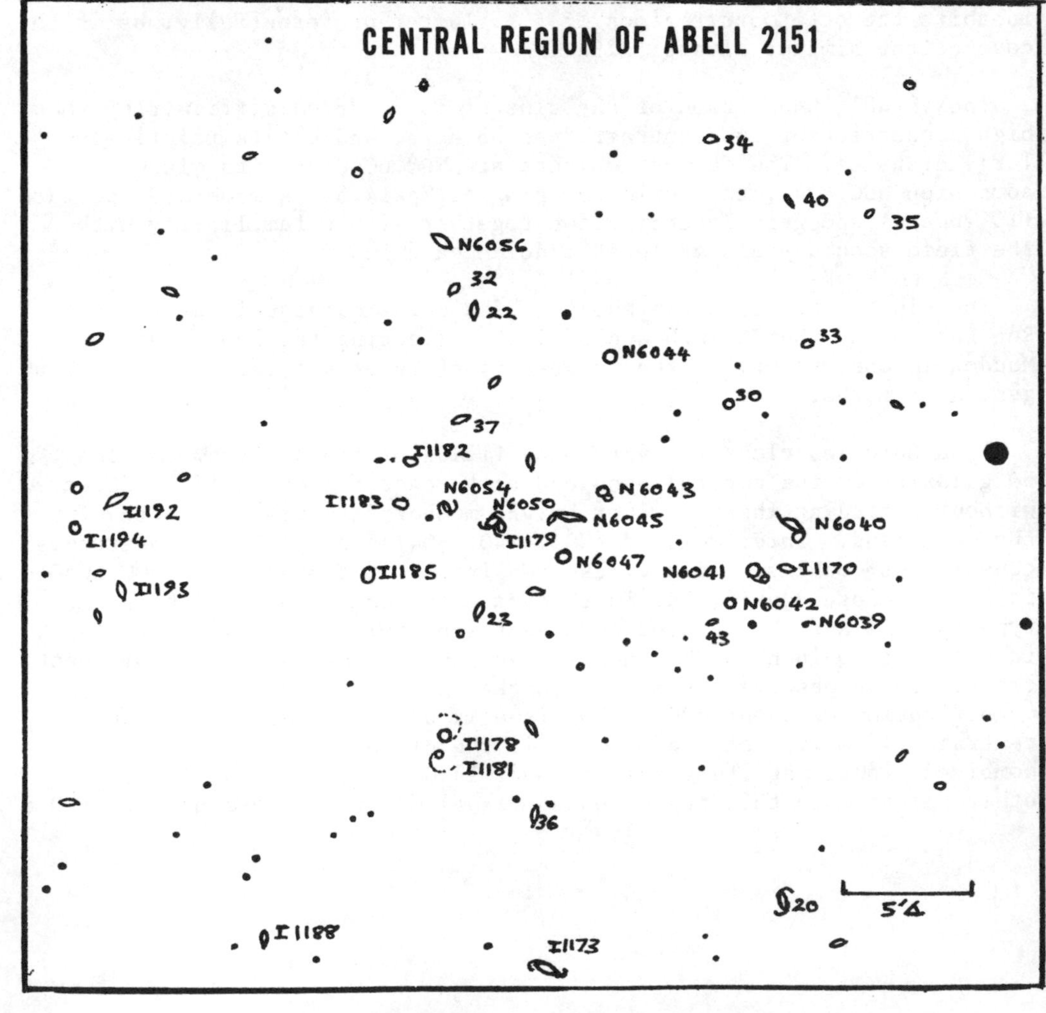

CENTRAL REGION OF ABELL 2151
34
40
35
N6056
32
22
33
N6044
30
37
I1182
N6054
N6043
I1192
I1183
N6050
N6045
N6040
I1194
I1179
N6047
N6041
I1170
I1185
I1193
N6042
23
N6039
43
I1178
I1181
36
20
5'
I1188
I1173

List of Catalogued Galaxies in A2151

The following data is based on a chart with V_{25} magnitudes published by Godwin and Peach[3] and the non-NGC/IC galaxies have been numbered following their convention. The NGC/IC identifications are due to Corwin[4], the latter paper containing comments on cross-identifications between various catalogues. The positions are those given in the MCG.

Chart	RA	Dec.	V_{25}	MCG	Type
35	01.9	18'07	15.5		
33	02.2	18'00'	15.5		
N 6040	02.25	17'53'	14.5	3-41-73	SAB(s)cdp
I 1170	02.28	17'51'	15		SBO
20	02.25	17'35'	14.5		
N 6041	02.4	17'52'	14.5	3-41-78	SABO
N 6042	02.45	17'51'	14.5	3-41-79	SAO
N 6044	02.7	18'02'	14.5	3-41-84	EO/SO
N 6043	02.8	17'54'	14.5	3-41-86	SAO
N 6047	02.9	17'52'	14.5	3-41-87	SA(s)Opec
N 6045	02.9	17'54'	14.5	3-41-88	SBcd:sp
IC 1173	02.95	17'34'	15.6p	3-41-93	SAB(s)C
IC 1179	03.13	17'53'	16	3-41-92	SAB(rs)d
IC 1178	03.3	17'44'	14.5	3-41-97	SAO
IC 1181	03.3	17'43'	15.0	3-41-98	SAB(S)O/a
N 6054	03.3	17'54'	15.0	3-41-99	(R')SB(s)b
22	03.2	18'03'	15.0		
32	03.3	18'04'	15.0		
N 6056	03.3	18'05'	14.5	3-41-100	SO/a?
37	03.3	17'58'	15.5		
N 6055	03.4	18'18'	14.5	3-41-101	SABO/a?
IC 1182	03.4	17'56'	14.5	3-41-104	SO†pec

Chart	RA	Dec.	V_{25}	MCG	Type
IC 1183	03.45	17'54'	14.5	3-41-103	SO pec
N 6057	03.4	18'19'	16	3-41-106	E1?
IC 1185	03.5	17'51'	14.5	3-41-110	SA(s)ab
IC 1188	03.9	17'36'	14.5		Sb
IC 1192	04.3	17'55'	NA		SB(s)0/a:sp
IC 1193	04.3	17'50'	NA		SBO? pec
IC 1194	04.4	17'53'	NA		E1 pec

NA = magnitude not available

Visual Observations

NGC.6034

(30) Considerably faint, very small with a faint nucleus.

NGC.6040

(82) Pair at 0.5' separation. Both pretty bright and well shown.

(30) Close double galaxy, pretty bright overall. The northern component (NGC.6040 A) is faint, pretty small and much extended, having an extremely faint nucleus in a bulge region. The southern component (NGC.6040 B) is faint, very small and round with a conspicuous nucleus.

(16) Small and slightly elongated with a PA of about -45°. Quite bright and brighter to the middle. NGC.6040B not observed. Brightest nebula in the cluster and first of a MP field of 3 galaxies.

NGC.6041

(107) Considerably bright double nebula. A very pretty pair.

(36) Pair at 0.3' separation. Immersed in a common envelope?

(30) Close double galaxy. The northern component, NGC.6041A is considerably faint, small and round with a conspicuous nucleus, whilst the southern component, NGC.6041 B is very faint, very small and round, possibly with a faint nucleus.

(16) Small and irregularly round. Quite bright and slowly brighter to the middle. Neither NGC.6041 B (attached SP) or companion (IC 1170) close preceding were observed.

NGC.6042

(30) Considerably faint, small and round with a stellar nucleus. In line with NGC.6040 and 6041.

(16) Almost stellar but quite bright. Third of three.

NGC.6043

(36) A very close double, the northern component being much the brighter. Both show faint nuclei.

(30) Faint, small and round.

(16) First of a MP field of 8 galaxies. Small and round. Faint but slowly brighter to the middle. A faint star lies close south following.

NGC.6044

(36) Considerably faint, small and round.

(16) Second of 8. Small and round, but quite bright and brighter to the middle. Faint star close preceding.

NGC.6045

(36) Companion on following end suspected.

(30) Faint, pretty large and very much extended with a very faint nucleus. One of the more distinctive members of the cluster.

(16) Third of 8. Small and slightly elongated E/W. One of the brightest in the field. Slowly brighter to the middle.

NGC.6047

(30) Faint, small and round. Pretty bright star north preceding.

(16) Fourth of 8. Almost stellar and faint but readily visible. Faint star NP.

NGC.6050 + IC 1179

(82) A close double galaxy at 0.3' separation. NGC.6050 appears pretty faint and diffuse with an irregularly extended, non-oval shape. Conspicuous nucleus and a very faint knot close north preceding. IC 1179 appears very faint and round, having a diffuse texture but no clearly distinguishable nucleus. Two much fainter galaxies near, are due south of IC 1179 and the other SP.

(16) Fifth of 8. Faint, irregularly round nebula slowly brightening to the middle. IC 1179 not separated if observed at all.

NGC.6054

(30) Very faint, very small with a bright star south following. (This object is faint compared to other NGC/IC cluster galaxies.)

NGC.6055

(30) Pretty faint, small and round having high surface brightness. Pretty faint star south preceding.

NGC.6056

(107) Considerably bright and irregularly extended with two anonymous companions SP, both bright. (Chart 32 and 22 - latter visible in a 16").

(30) Pretty faint, small and round.

(16) Small and slightly elongated in PA of about -45°. Quite bright and brighter to the middle. In a field of 3 galaxies with IC 1182 and Chart (22).

NGC.6057

(30) Faint, very small and round with a very faint nucleus.

NGC.6061

(30) Pretty faint, small and round having a conspicuous nucleus.

IC 1170

(107) Pretty bright, much extended with a faint star close SF.

(36) As 30 inch observation but with an extremely faint star close SF.

(30) Very faint, extremely small and round. Close preceding NGC.6041.

IC 1173
(36) Faint and diffuse. Clearly a spiral.

(30) Very faint and round, having a diffuse texture.

IC 1181 + IC 1178
(30) IC 1178 appears pretty faint and small having a conspicuous nucleus. IC 1181 is faint, very small and round. A pretty pair at 0.5' separation.

IC 1182
(30) Faint, small and round. Somewhat diffuse with a faint nucleus.

(16) Seventh of 8. Very faint, irregularly round and almost stellar. Star P.

IC 1183
(30) Faint, small and round with a nucleus. An extremely faint star lies about 0.4' south preceding. (Possibly NGC.6054.)

(16) Sixth of 8. Similar to IC 1182. Star preceding.

IC 1185
(30) Faint, small and round, having a conspicuous nucleus. (Possibly a foreground star according to Corwin.)

IC 1186
(30) Faint, small and possibly a little extended. Many field stars close east.

IC 1188
(36) A very close double galaxy, both nuclei being visible less than 15" apart. Preceding component is distinctly brighter than the following.

(30) Very faint, very small and round.

IC 1189
(30) Faint and much extended, having a conspicuous nucleus. A very faint companion close S visible in 36 inch.

IC 1191
(30) Extremely faint, very small and round having a star very close following. (Actually a double galaxy but following component not noticed to be non-stellar.)

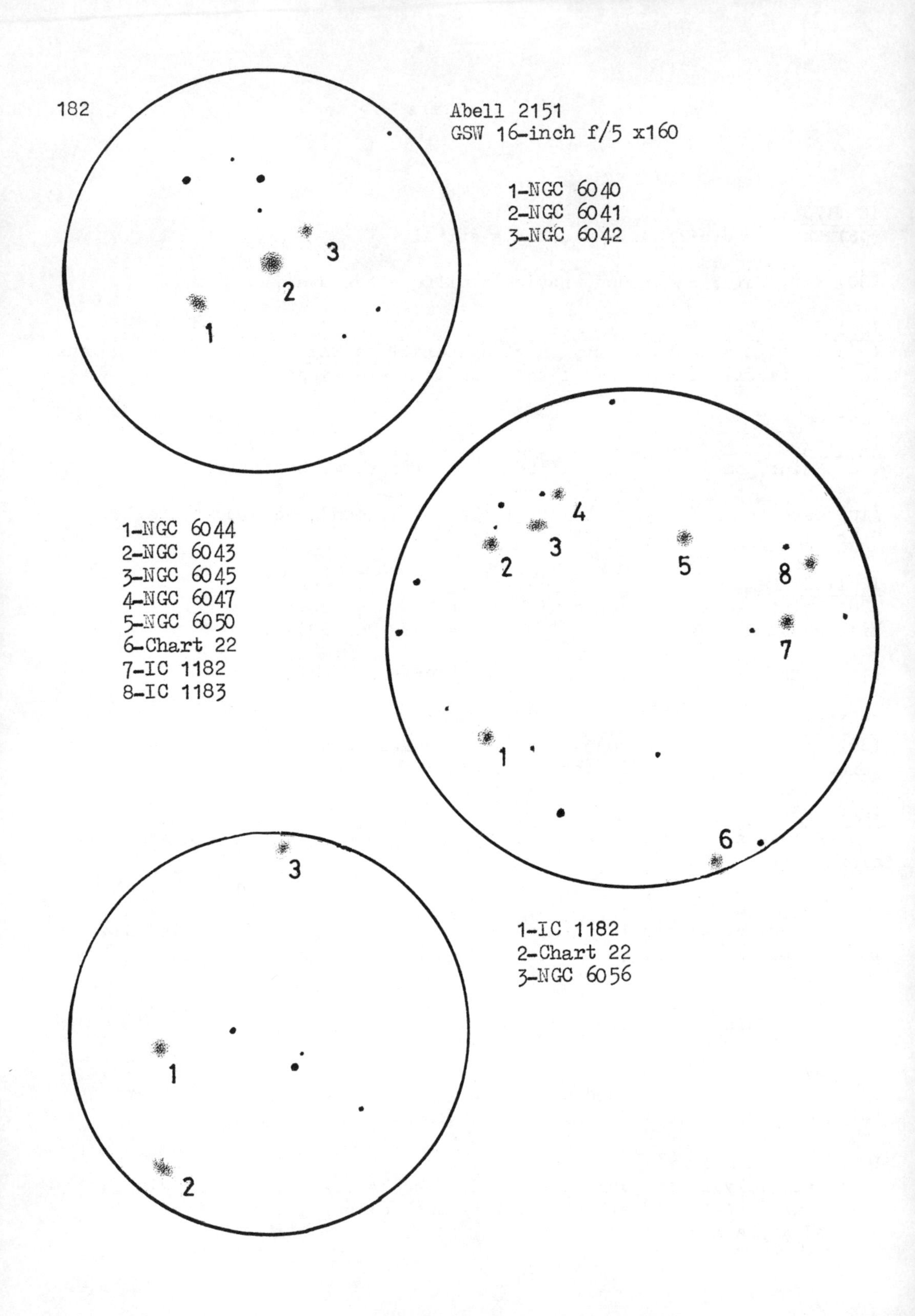
Abell 2151
GSW 16-inch f/5 x160
1-NGC 6040
2-NGC 6041
3-NGC 6042
1-NGC 6044
2-NGC 6043
3-NGC 6045
4-NGC 6047
5-NGC 6050
6-Chart 22
7-IC 1182
8-IC 1183
1-IC 1182
2-Chart 22
3-NGC 6056

IC 1192
(30) Very faint, small and round. First of a subgroup east of the central region.

IC 1193
(30) Faint, small and round with a nucleus. Two anonymous galaxies faintly visible close north following and south following.

IC 1194
(30) Considerably faint, small and round with a conspicuous nucleus. Brightest in the eastern subgroup.

IC 1195
(30) Considerably faint, small and diffuse. (Faint nucleus suspected using 36 inch.)

14. Abell 2197

Abell	RA	Dec.	AD	AR	Z	RS
A2197	$16^h26.5$	$+41^o1$	1	1	0.303	L

Abell 2197 is a splendid cluster lying to the north west of the keystone of Hercules at a distance of about 180 Mpc. It is dominated by the giant galaxies NGC.6173, NGC.6160 and NGC.6146, the first galaxy having a gigantic envelope characteristic of cD galaxies(1). These three galaxies lie on a line about a degree in length, along which lie many minor galaxies. The cluster is not as linearly concentrated as A426, however, since a considerable number of quite large galaxies lie off it. Thompson(2) has detected a significant degree of alignment of galactic major axes in the cluster at PA 90°. Although his histogram is quite markedly peaked at this angle, the effect is not at all apparent on casual inspection of a 16 inch × 20 inch enlargement of a 48 inch Schmidt plate of the cluster, from which the chart below was prepared. Interestingly, the above alignment is approximately along the direction of the NGC.6173 - NGC.6146 chain. Although A2197 contains a cD galaxy, it is not a typical cD cluster. For example, although cD clusters are almost always aligned along the axis of the cD galaxy, A2197 and NGC.6173 are not at all aligned. One of the most interesting galaxies in the cluster is NGC.6175 which is probably a line-of-sight juxtaposition of an E0 galaxy, with an edge-on S0 situated to the north. The justaposition is not apparent visually in moderate apertures.

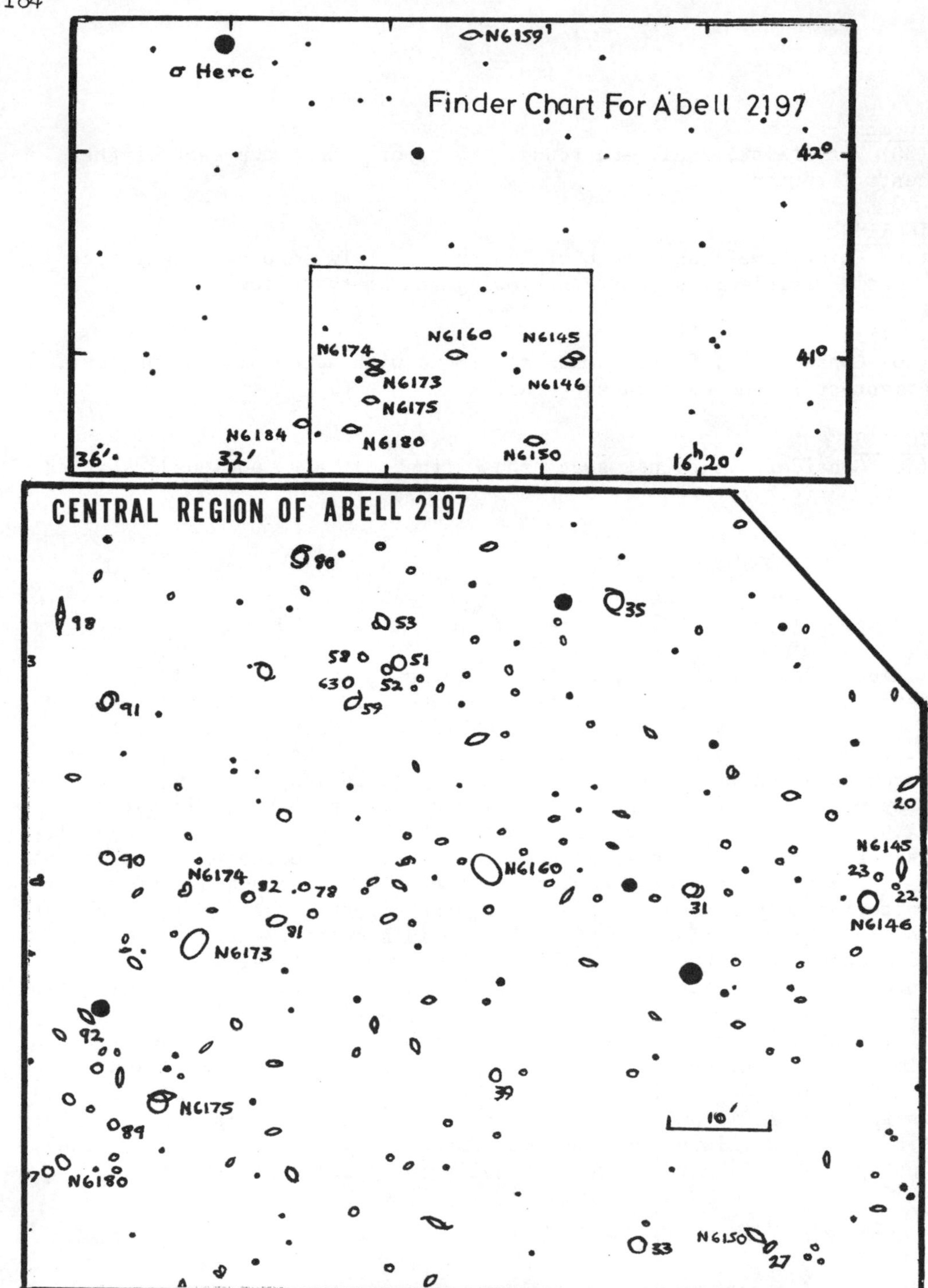

Finder Chart For Abell 2197
σ Herc
N6159
N6160
N6145
N6174
N6173
N6146
N6175
N6184
N6180
N6150
42°
41°
36'
32'
16h20'
CENTRAL REGION OF ABELL 2197
N6160
N6174
N6173
N6175
N6180
N6145
N6146
N6150
10'

Although A2197 contains a liberal sprinkling of obvious spiral galaxies, most of the main galaxies appear to be elliptical or lenticular. NGC.6146 is the brightest galaxy visually, followed by NGC.6173 and NGC.6160. All three should be visible in a 12 inch short focus reflector. However, most of the other galaxies are relatively difficult requiring more than 12 inches for detection.

List of Catalogued Galaxies

The following data is based entirely upon the MCG.

MCG	RA	Dec.	Mp	NGC/IC
7-34-20	$16^h23.3'$	$+41^o08'$	15	
7-34-21	23.4'	$41^o02'$	14.6	NGC.6145
7-34-22	23.4'	$41^o00'$	17	
7-34-23	23.45'	$41^o01'$	14.7	
7-34-24	23.55'	$40^o59'$	14	NGC.6146
7-34-27	24.1'	$40^o33'$	15.2	
7-34-28	24.1'	$40^o26'$	15	
7-34-29	24.2'	$40^o34'$	14.7	NGC.6150
7-34-31	24.75'	$41^o00'$	15	
7-34-33	25.0'	$40^o34'$	14	
7-34-35	25.25'	$41^o21'$	13	
7-34-39	25.95'	$40^o46'$	15	
7-34-42	26.05'	$41^o00'$	14	NGC.6160
7-34-47	26.6'	$40^o24'$	14	
7-34-51	26.8'	$41^o15'$	15	
7-34-52	26.85'	$41^o14'$	15	
7-34-53	26.85'	$41^o18'$	14	
7-34-58	26.9'	$41^o15'$	16	
7-34-59	26.9'	$41^o11'$	16	
7-34-63	26.95'	$41^o13'$	16	
7-34-78	27.25'	$40^o57'$	16	

MCG	RA	Dec.	Mp	NGC/IC
7-34-81	$16^h27.5'$	$+40^o56'$	15	
7-34-82	27.27'	$40^o57'$	15	
7-34-83	28.1'	$40^o53'$	14.2	NGC.6173
7-34-85	28.1'	$40^o57'$	15	NGC.6174
7-34-87	28.3'	$40^o42'$	14	NGC.6175
7-34-89	28.65'	$40^o41'$	15	
7-34-90	28.7'	$41^o00'$	14	
7-34-91	28.7'	$41^o11'$	15	
7-34-95	28.9'	$40^o37'$	15	NGC.6180
7-34-97	29.0'	$40^o36'$	15	
7-34-98	29.0'	$41^o17'$	16	
7-34-103	29.4'	$41^o14'$	14	

Visual Observations

A 16" f/5 Newtonian reflector was used to sweep up the following cluster galaxies.

NGC.6145 Small oval about 30" major axis oriented N/S. Quite bright and brighter to the middle.

NGC.6146 Round, being about 30" in diameter. Bright and much brighter to the middle. Star mag. 10 or 11 following by about 30".

NGC.6160 Oval about 45" by 30". Quite bright and slowly brighter to the middle to a faint stellar nucleus. Faint star in contact on NF tip. Nebula oriented in PA -45°.

NGC.6174 Small and round, almost stellar. Quite bright and brighter to the middle.

NGC.6173 Oval of similar size to NGC.6160, but brighter overall. Quite bright and brighter to the middle.

NGC.6175 Small, irregularly round, Quite bright and brighter to the middle. Between two stars.

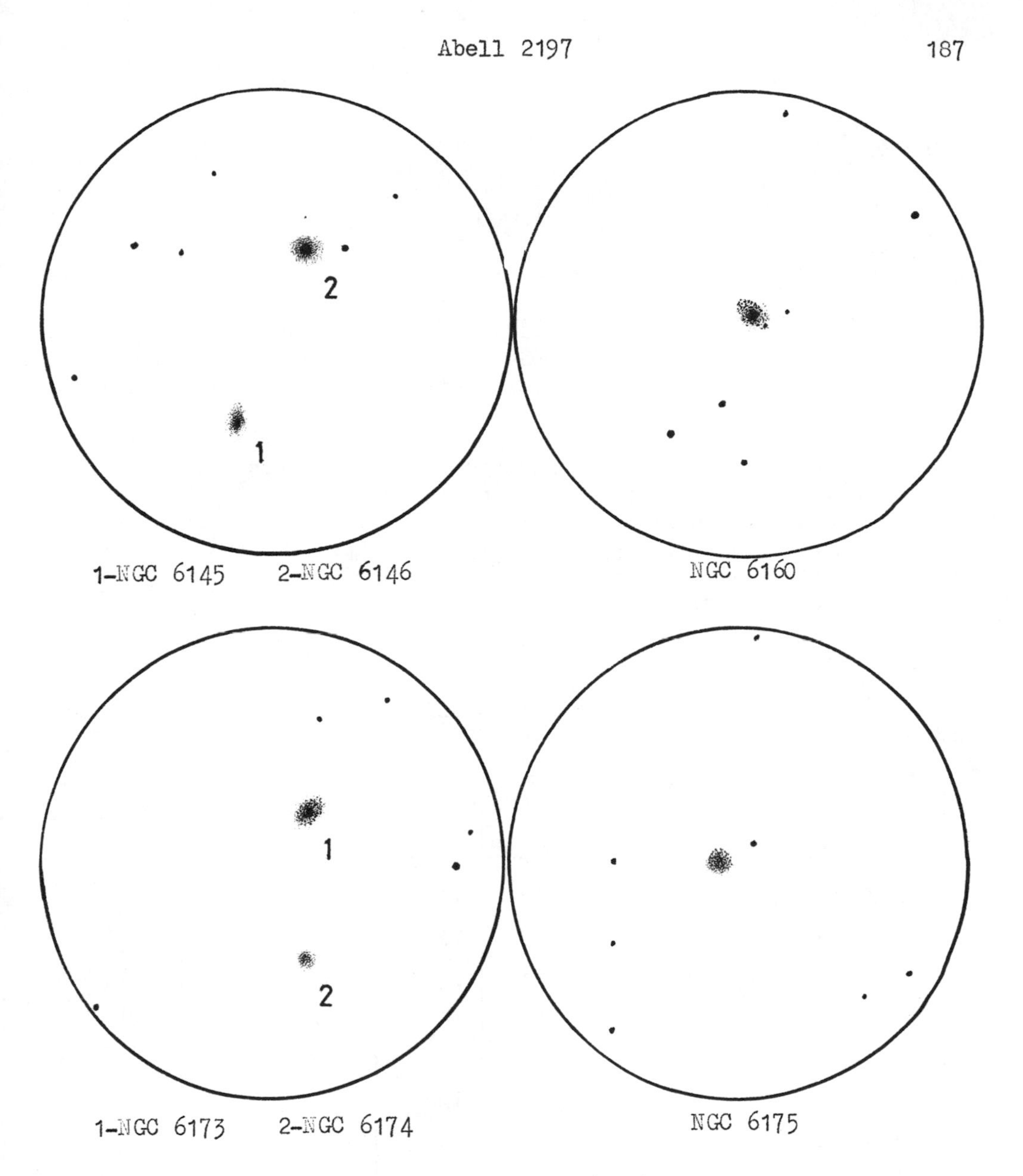

GSW 16-inch f/5 x160 (12')

15. Abell 2199

WS	Abell	RA	Dec.	AD	AR	Z	RS
	2199	$16^h26.9$	$+39^o38'$	1	2	0.0306	cD

Abell 2199[1] is probably the archetypal cD cluster, being dominated by the gigantic galaxy NGC.6166 whose halo has been traced to a diameter of about 1 Mpc[2] (19' at 184 Mpc). Refering to the chart, it is clear that most of the core of the cluster is contained within the halo, and distributed roughly about the direction of its major axis.

NGC.6166 is an easy visual object, having photographic magnitude 13.3, and in good skies with a moderate aperture, its field is rich in galaxies Most of the visually observable galaxies (with the notable exception of NGC.6158) are to be found north following the core. The cluster is only about 1.5^o away from A2197 on the sky (about 5 Mpc) and their relative distance is very small, ($\Delta z = 3 \times 10^{-4}$). Both clusters probably form part of a supercluster.

List of Catalogued Galaxies

The data in the following table is based upon the MCG and is cross-correlated with the RNGC.

MCG	RA	Dec.	Mp	RNGC
7-34-25	$16^h23.75'$	$+39^o52'$	14	
7-34-26	24.0'	$39^o40'$	15	
7-34-36	25.7'	$39^o11'$	15	
7-34-41	26.0'	$39^o27'$	14.7	NGC.6158
7-34-43	26.2'	$39^o20'$	15	
7-34-44	26.2'	$39^o21'$	15	
7-34-45	26.25'	$39^o23'$	15	
7-34-46	26.5'	$39^o54'$	14	
7-34-48	26.7'	$39^o40'$	15.1	RNGC 6166C
7-34-49	26.7'	$39^o42'$	16	
7-34-50	26.8'	$39^o37'$	16	RNGC 6166A
7-34-54	26.85'	$39^o38'$	16	RNGC 6166D
7-34-55	26.9'	$39^o37'$	16	

MCG	RA	Dec.	Mp	RNGC
7-34-56	$16^h26.9'$	$39^o36'$	14.9	
7-34-57	26.9'	$39^o09'$	15	
7-34-60	26.9'	$39^o38'$	13.30	NGC.6166
7-34-64	27.0'	$39^o34'$	15	
7-34-65	27.0'	$39^o34'$	16	
7-34-66	27.0'	$39^o36'$	15.3	STAR
7-34-67	27.0'	$39^o31'$	15	
7-34-68	27.0'	$39^o44'$	16	
7-34-69	27.0'	$39^o33'$	15	
7-34-70	27.0'	$39^o37'$	16	
7-34-71	27.05'	$39^o42'$	16	
7-34-72	27.05'	$39^o45'$	16	
7-34-73	27.1'	$39^o44'$	16	
7-34-74	27.1'	$39^o42'$	16	
7-34-75	27.1'	$39^o55'$	15	
7-34-76	27.15'	$39^o39'$	15.0	RNGC 6166B
7-34-79	27.4'	$39^o35'$	15	
7-34-86	28.15'	$39^o50'$	14	
7-34-94	28.9'	$39^o54'$	15	
7-34-99	29.0'	$39^o16'$	16	
7-34-100	29.1'	$39^o17'$	15	
7-34-102	29.25'	$39^o55'$	14.7	
7-34-104	29.4'	$39^o51'$	14	
7-34-105	29.4'	$39^o54'$	15	
7-34-107	29.6'	$39^o13'$	16	
7-34-108	29.65'	$39^o14'$	16	

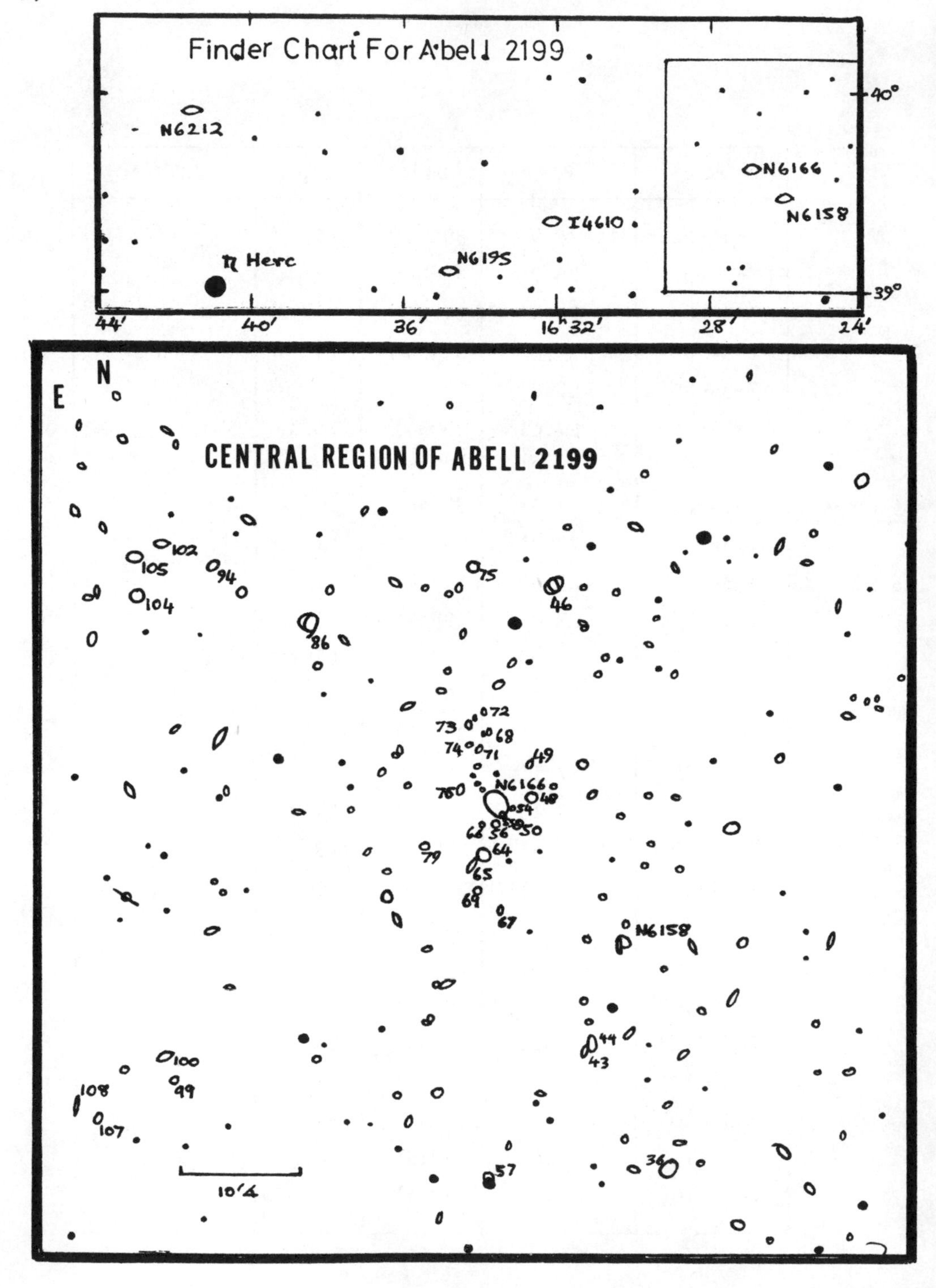
Finder Chart For Abell 2199
N6212
η Herc
N6195
I4610
N6166
N6158
40°
39°
44'
40'
36'
16h32'
28'
24'
E
N
CENTRAL REGION OF ABELL 2199
102
105
94
104
75
46
86
72
73
68
74
71
49
N6166
76
48
54
58
63
56
50
64
79
65
69
67
N6158
44
43
100
99
108
107
57
36
10'4

Comments on Identifications

Figure (10) is a sketch of the field of NGC.6166 based on a 5 m photograph published in 'Scientific American'[3] and an elargement of a 48 inch Schmidt plate. Figure (11) plots the MCG and RNGC galaxies in a similar field in an attempt to cross-correlate them. Because of the fact that catalogue positions are rounded off to the nearest tenth of a minute of RA and nearest minute of arc, identification of galaxies in this crowded field is not immediately obvious. The most ambiguous identification is for MCG 7-34-66, which could either be the indicated star or the lenticular galaxy immediately north of MCG 7-34-64. Again RNGC 6166A could be MCG 7-34-50 or 56. The former is preferred since the latter is virtually stellar even on the 200 inch plate.

Visual Observations

The following visual observations were made with 30 inch and 16 inch reflectors. Most of the cluster galaxies are very faint, requiring moderate apertures but the brightest galaxy, NGC.6166 should be visible in a 10 inch.

MCG 7-34-48 (RNGC 6166C)
(30) Faint, very small and round, having a stellar nucleus.

(16½) North preceding NGC.6166 with a star close south following. A nebulous knot - definitely a galaxy.

(16) Very small, only just non-stellar. Faint, but the brightest of the companions of NGC.6166.

MCG 7-34-49
(30) Stellar, a faint star lies close south following. (Probably a galaxy (GSW)).

MCG 7-34-50 (RNGC 6166A)
(30) Faint, very small and round with a stellar nucleus.

(16½) South preceding NGC.6166 with a faint star preceding. Easily visible showing an irregularly round nebula with a slight central brightening.

(16) Very faint, almost stellar.

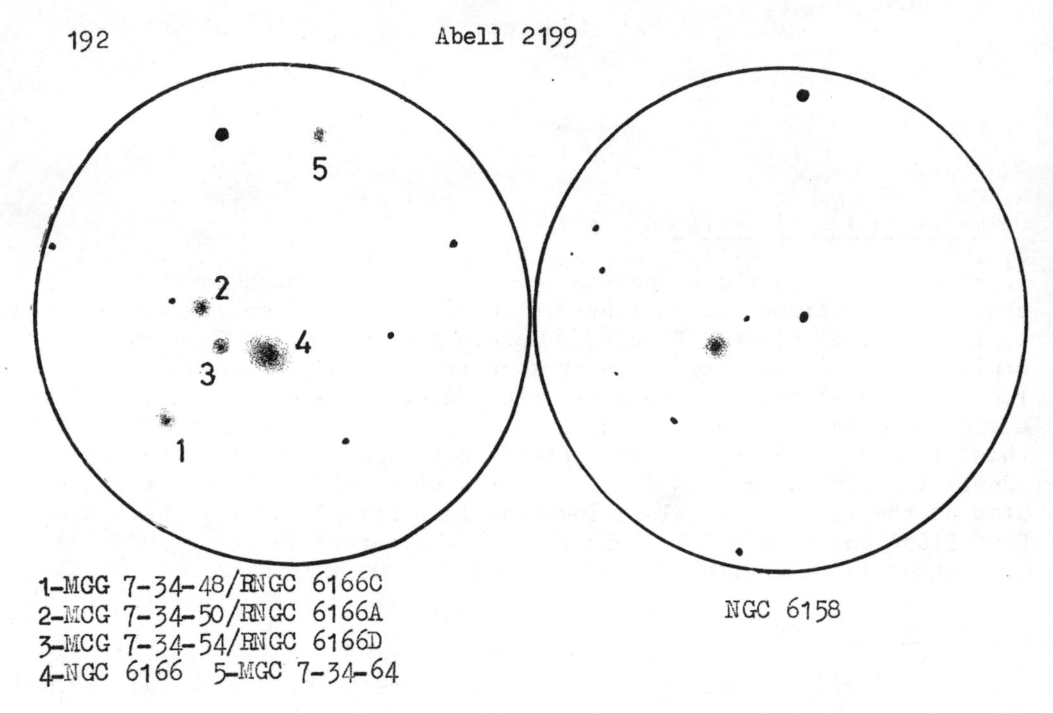
5
2
4
3
1
1-MGG 7-34-48/RNGC 6166C
2-MCG 7-34-50/RNGC 6166A
3-MCG 7-34-54/RNGC 6166D
4-NGC 6166 5-MGC 7-34-64
NGC 6158

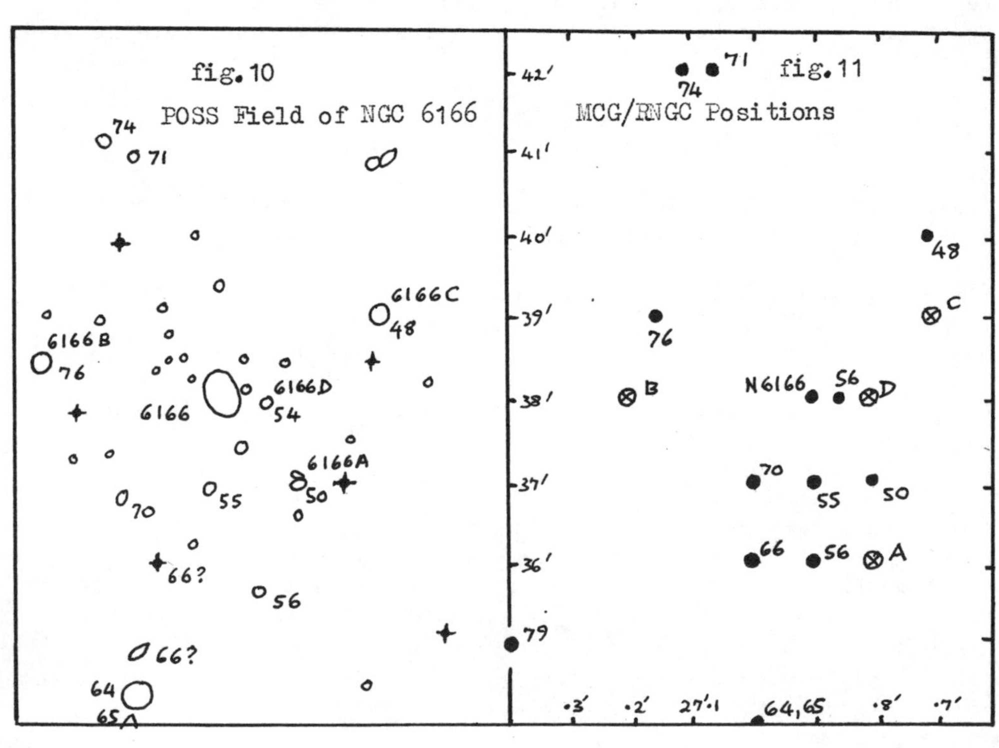
fig. 10
POSS Field of NGC 6166
74
71
6166C
48
6166B
76
6166
6166D
54
6166A
50
70
55
66?
56
66?
64
65A
fig. 11
MCG/RNGC Positions
42'
41'
40'
39'
38'
37'
36'
74
71
48
C
76
B
N6166
56
D
70
55
50
66
56
A
79
•3'
•2'
27'•1
64,65
•8'
•7'

MCG 7-34-54 (RNGC 6166D)
(30) Extremely faint, extremely small and round. 0.6' preceding NGC.6166.

(16) Almost stellar object preceding NGC.6166. Extremely difficult.

MCG 7-34-55
(30) Very faint, very small round nebula with a stellar nucleus.

(16½) Almost due south of NGC.6166 and between it and MCG 7-34-65. Faint and almost stellar.

MCG 7-34-64
(30) Faint, very small and round. Somewhat diffuse. 7-34-65. Stellar close south.

(16½) South and slightly following NGC.6166. Faint, irregularly round and of uniform surface brightness.

(16) Small and round. Faint but slightly brighter to the middle.

NGC.6166

(30) Pretty bright, considerably small oval with two nuclei and some mottling between them.

(16½) Irregularly round and of medium size brightening slightly in the middle.

(16) Oval with major axis of about 1' aligned with a PA of about 30°. Quite bright and slowly brighter to the middle.

MCG 7-34-66?
(30) Stellar.

MCG 7-34-70
(30) Stellar. Anon stellar galaxy close north following.

MCG 7-34-76 (RNGC 6166B)
(30) Stellar. (Stellar object preceding and NF a star is probably 7-34-71 (GSW).)

NGC.6158
(16) Round, having a diameter of about 30". Quite bright and brighter to the middle. Faint star preceding by 30" of arc.

APPENDIX: CATALOGUE OF ABELL CLUSTERS WITH D $\leqslant$ 3

The following catalogue contains all Abell clusters at galaxies in distance groups 0-3 with declination north of -25° (1855). The positions were kindly precessed from 1855 to 1950 by R.W. Argyle. Data columns contain respectively Abell catalogue number, a pair comprising (distance class, richness class), RA and Dec. for 1950.

Abell	(AD, AR)	RA (1950)	Dec. (1950)
0071	3, 0	$00^{h}35.1'$	$29^{o}19'$
0076	3, 0	$00^{h}37.2'$	$05^{o}40'$
0102	3, 1	$00^{h}46.1'$	$01^{o}06'$
0119	3, 0	$00^{h}53.8'$	$-01^{o}32'$
0147	3, 0	$01^{h}05.6'$	$01^{o}55'$
0151	3, 1	$01^{h}06.4'$	$-15^{o}42'$
0154	3, 1	$01^{h}08.3'$	$-17^{o}24'$
0168	3, 2	$01^{h}12.6'$	$-00^{o}01'$
0179	3, 0	$01^{h}19.1'$	$19^{o}14'$
0194	1, 0	$01^{h}23.0'$	$-01^{o}46'$
0195	3, 0	$01^{h}24.2'$	$18^{o}55'$
0240	3, 0	$01^{h}39.3'$	$07^{o}23'$
0262	1, 0	$01^{h}49.9'$	$35^{o}54'$
0277	3, 1	$01^{h}53.3'$	$-07^{o}38'$
0278	3, 0	$01^{h}54.4'$	$31^{o}59'$
0347	1, 0	$02^{h}22.7'$	$41^{o}39'$
0338	3, 0	$02^{h}27.7'$	$13^{o}25'$
0376	3, 0	$02^{h}42.6'$	$36^{o}39'$
0397	3, 0	$02^{h}54.1'$	$15^{o}45'$
0400	1, 1	$02^{h}55.0'$	$05^{o}50'$
0401	3, 2	$02^{h}56.2'$	$13^{o}23'$
0426	0, 2	$03^{h}15.3'$	$41^{o}20'$
0496	3, 1	$04^{h}31.3'$	$-13^{o}22'$
0505	3, 0	$04^{h}51.6'$	$79^{o}56'$

Abell	(AD, AR)	RA (1950)	Dec. (1950)
0514	3, 1	$04^h45.5'$	$-20^o32'$
0539	2, 1	$05^h13.9'$	$06^o25'$
0553	3, 0	$06^h08.8'$	$48^o37'$
0568	3, 0	$07^h04.3'$	$35^o08'$
0569	1, 0	$07^h05.4$	$48^o42'$
0576	2, 1	$07^h17.4'$	$55^o50'$
0592	3, 1	$07^h39.9'$	$09^o30'$
0595	3, 0	$07^h45.1'$	$52^o12'$
0634	3, 0	$08^h10.5'$	$58^o12'$
0671	3, 0	$08^h25.4'$	$30^o36'$
0754	3, 2	$09^h06.5'$	$-09^o27'$
0757	3, 0	$09^h09.4'$	$47^o56'$
0779	1, 0	$09^h16.9'$	$33^o59'$
0838	3, 0	$09^h34.6'$	$-04^o47'$
0978	3, 1	$10^h18.0'$	$-06^o17'$
0979	3, 0	$10^h18.0'$	$-07^o39'$
0993	3, 0	$10^h19.4'$	$-04^o43'$
0999	3, 0	$10^h20.7'$	$13^o06'$
1016	3, 0	$10^h24.5'$	$11^o14'$
1035	3, 2	$10^h29.2'$	$40^o29'$
1069	3, 0	$10^h37.4'$	$-08^o22'$
1139	3, 0	$10^h35.5'$	$01^o47'$
1142	3, 0	$10^h58.3'$	$10^o49'$
1185	2, 1	$11^h08.1'$	$28^o57'$
1187	3, 1	$11^h08.9'$	$39^o51'$
1213	2, 1	$11^h13.8'$	$29^o33'$
1225	3, 1	$11^h18.5'$	$54^o03'$
1228	1, 1	$11^h18.9'$	$34^o37'$
1254	3, 1	$11^h23.9'$	$71^o22'$
1257	3, 0	$11^h23.4'$	$35^o37'$

Abell	(AD, AR)	RA (1950)	Dec. (1950)
1267	3, 0	$11^h25.3'$	$27^o09'$
1270	3, 0	$11^h26.7'$	$54^o21'$
1291	3, 1	$11^h29.3'$	$56^o19'$
1314	1, 0	$11^h32.1'$	$49^o20'$
1318	3, 1	$11^h33.7'$	$55^o15'$
1367	1, 2	$11^h41.9'$	$20^o07'$
1377	3, 1	$11^h44.3'$	$56^o01'$
1436	3, 1	$11^h58.0'$	$56^o32'$
1500	3, 0	$12^h11.5'$	$75^o40'$
1631	3, 0	$12^h50.2'$	$-15^o10'$
1656	1, 2	$12^h57.4'$	$28^o15'$
1691	3, 1	$13^h09.5'$	$39^o27'$
1773	3, 1	$13^h39.6'$	$02^o30'$
1781	3, 0	$13^h42.2'$	$30^o06'$
1800	3, 0	$13^h47.3'$	$28^o20'$
1831	3, 1	$13^h56.9'$	$28^o14'$
1890	3, 0	$14^h15.0'$	$08^o25'$
1904	3, 2	$14^h20.3'$	$48^o48'$
1983	3, 1	$14^h50.4'$	$16^o57'$
1991	3, 1	$14^h51.9'$	$19^o18'$
2022	3, 1	$15^h02.2'$	$28^o38'$
2052	3, 0	$15^h14.2'$	$07^o12'$
2063	3, 1	$15^h20.6'$	$08^o49'$
2065	3, 2	$15^h20.6'$	$27^o54'$
2079	3, 1	$15^h26.0'$	$29^o03'$
2124	3, 1	$15^h43.0'$	$36^o14'$
2147	1, 1	$16^h06.0'$	$16^o03'$
2148	3, 0	$16^h01.2'$	$25^o36'$
2151	1, 2	$16^h02.9'$	$17^o53'$

Abell	(AD, AR)	RA (1950)	Dec. (1950)
2152	1, 1	$16^h03.1'$	$16^o35'$
2162	1, 0	$16^h10.5'$	$29^o40'$
2197	1, 1	$16^h26.5'$	$41^o01'$
2199	1, 2	$16^h26.8'$	$39^o38'$
2241	3, 0	$16^h57.8'$	$81^o39'$
2248	3, 0	$16^h59.9'$	$77^o05'$
2249	3, 0	$17^h07.9'$	$34^o32'$
2255	3, 2	$17^h12.2'$	$64^o09'$
2256	3, 2	$17^h06.6'$	$78^o47'$
2319	3, 1	$19^h19.2'$	$43^o53'$
2399	3, 1	$21^h54.9'$	$-08^o21'$
2572	3, 0	$23^h15.9'$	$18^o28'$
2589	3, 0	$23^h21.4'$	16^o33
2593	3, 0	$23^h22.0'$	$14^o22'$
2625	3, 0	$23^h33.8'$	$20^o15'$
2626	3, 0	$23^h34.0'$	$20^o53'$
2630	3, 0	$23^h35.0'$	$15^o34'$
2634	1, 1	$23^h35.9'$	$26^o46'$
2657	3, 1	$23^h42.3'$	$08^o53'$
2666	1, 0	$23^h48.4'$	$26^o53'$

INTRODUCTION

The following catalogue of 15 groups of galaxies was prepared from visual observations made by Webb Society members using telescopes of large and moderately large aperture by amateur standards, and on the whole, consists of observations of very faint galaxies. However, some of the brighter members of these groups should be visible to small apertures in good conditions, and the fainter members should present an interesting challenge to observational ability.

For the purposes of the catalogue, a 'group' is defined as a poor cluster having at least four galaxies brighter than the 15th magnitude, including at least one NGC or IC object in a circular area of sky at most $\frac{1}{2}^{\circ}$ in diameter. Thus at least four galaxies should be visible in the LP field of a sufficiently powerful telescope. Some of the groups selected for the catalogue are classed as 'clusters', that is, they contain about a score of galaxies in a compact area. The NGC.5416 group is one of these. The distinction between a 'poor cluster' and a 'group' is one of convenience in that in this volume, 'cluster' is reserved for objects in Abell's catalogue which have well-defined population and compactness criteria.

The main catalogue consists of a collection of verbal descriptions of objects in the groups tabled below, and many of these are supplemented by field drawings collected together at the end of the catalogue.

A group is named after the brightest NGC galaxy within the group. If two or more galaxies are assigned the same magnitude, the group is named after the most preceding NGC galaxy of minimal magnitude. The position of the dominant galaxy is also used as the position of the group on the sky. Each description is headed by four parameters:- (1) the name of the group, (2) the RA and (3) the Declination, (1950) and (5) nG - the number of galaxies of magnitude 16 or less within a square box of side 30' of arc recorded in the MCG.

Group	Observers
48	MJT,GSW
68	GSW
80	GSW,RB
128	GSW,RB
383	GSW
507	GSW,MJT
529	MJT

Group	Observers
973	MJT,GSW
3395	MJT,GSW
4005	MJT,GSW,RB
5353	MJT,GSW
5416	GSW
7320	GSW,JB,ESB

Catalogue of Groups of Galaxies

NGC	RA	Dec.	nG
48	$23^h11.3$	$+47^o58'$	6

MCG	RA	Dec.	Mp	NGC/IC
8-1-28	$23^h11.0'$	$47^o52'$	15	IC 1534
8-1-30	11.2'	$47^o52'$	15	IC 1535
8-1-31	11.3'	$47^o58'$	13	NGC.48
8-1-32	11.6'	$47^o51'$	15	IC 1536
8-1-33	11.7'	$47^o58'$	13	NGC.49
8-1-35	11.9'	$47^o59'$	13	NGC.51

IC 1534
(16½) Small, very faint and irregularly round. Brightens to the centre to developed core. Star close north following.

IC 1535
(16½) Directly following the star NF IC 1534. Faint, but larger than IC 1534. Extended being slightly NP/SF (170°). Uniform surface brightness.

IC 1536
(16½) Irregularly round, of medium size, and brightening to a small nucleus seeming eccentric to the southern end at high power. Between two faint stars.

NGC.48
(16½) First of three in line south of the three IC galaxies. Medium size, faint but easily seen being irregularly round and brightening a little in the middle. A star follows the galaxy.

(16) First of 3. Quite large and of low surface brightness. Slightly elongated approximately E/W. Diffuse and slightly brighter to the middle.

NGC.49

(16½) Smaller than NGC.48 and irregularly round with a small central brightening. The envelope is diffuse. Double star south following.

(16) Quite faint and almost stellar.

NGC.51

(16½) The third in line and the brightest, appearing quite large and irregular in shape. The central area is bright. A faint star lies on the southern edge.

(16) Round and brighter to the middle. Brightest of the three galaxies.

NGC	RA	Dec.	nG
68	$00^h15.7'$	$+29^o47'$	8

MCG	RA	Dec.	Mp	NGC/IC
5-1-64	$00^h15.65$	$29^o46'$	15.3	67
5-1-65	15.7	$29^o47'$	14.6	68
5-1-66	15.7	$29^o45'$	15.9	69
5-1-67	15.75	$29^o48'$	15.3	70
5-1-68	15.8	$29^o47'$	14.8	71
5-1-69	15.85	$29^o45'$	14.7	72
5-1-70	16.0	$29^o46'$	15.9?	
5-1-71	16.25	$29^o47'$	16	74

The NGC 68 group is an extremely compact group of almost stellar galaxies in Andromeda. It is also known as ZW 499-13. The field of the group is confused by numerous faint stars, and because visual observation is difficult, a chart of this group is included below, having a scale of 16' × 16', taken from a POSS print.

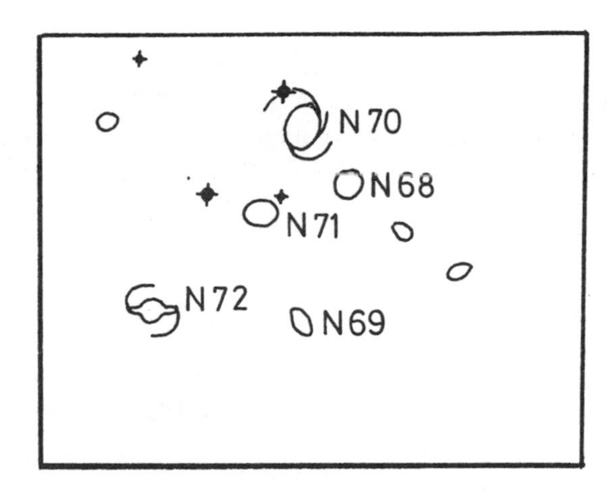

The identification of MCG 5-1-70 with a galaxy is rather difficult. The group has been observed with a 16 inch reflector, and a drawing of the group showing three or four of the objects is included in the album following this catalogue. The main knot of galaxies appears as a hazy nebulous area, included in which are several faint stellar objects, either stars or galactic nuclei. Only NGC.72 appears separate.

NGC	RA	Dec.	nG
80	$00^h18.6$	$+22^o05'$	9

MCG	RA	Dec.	Mp	NGC/IC
4-2-1	$00^h18.0$	$+22^o20'$	15	IC 1542
4-2-3	18.4	$22^o18'$	14	NGC.79
4-2-4	18.6	$22^o05'$	13.9	NGC.80
4-2-5	18.7	$22^o10'$	14.3	NGC.82/83
4-2-7	18.8	$22^o14'$	15	NGC.85a
4-2-7	18.9	$22^o14'$	15	NGC.85b
4-2-9	18.9	$22^o17'$	16	NGC.86
4-2-11	19.2	$22^o08'$	14	NGC.91
4-2-12	19.4	$22^o09'$	14	NGC.93
4-2-13	19.4	$22^o27'$	15	
4-2-14	19.7	$22^o17'$	17	NGC.46

The NGC.80 group is also known as Zw 478.5, and is situated approximately ½° following a point halfway between alpha-Andromeda and gamma-Pegasus. The following seven galaxies were observed using 30 inch and 16 inch reflectors.

IC 1542
(30) Considerably faint, small and diffuse. Forms a triangle with two faint stars to the south.

NGC.79
(30) Pretty bright, very small and round, having a conspicuous nucleus.

(16) Very faint indistinct blur. Possibly a faint double star or a star may be involved with the galaxy.

NGC.80
(30) Bright, small and round with a mottled texture.

(16) Brightest in a medium power field of three galaxies. Round (45") and quite bright being much brighter to the middle.

NGC.83
(30) Bright, small and round. Three pretty bright stars following.

(16) Small oval nebula with a major axis of about 45". Slightly elongated approximately E/W. Quite bright and slowly brighter to the middle. Three stars of mag. about 10 following.

NGC.84
(30) Very faint, small and round. Diffuse. Two stars north preceding.

NGC.85
(30) Pretty faint, small and round, having a stellar nucleus. Forms a double with IC 1546. Following by 0.8' of arc. (NGC.85b).

(16) Very faint, irregular nebulosity with two faint stars involved. Possible the nuclei of NGC.85a and b.

NGC.86
(30) Pretty faint, small ellipse. Mottled. A pretty bright star very close south.

IC .546 (NGC.85b)
(30) Pretty faint, small much extended nebula close south following NGC.85.

NGC.91
(30) Faint, very small and round.

(16) Small and slightly extended E/W. Star south preceding. Quite bright.

NGC.93
(30) Pretty bright, pretty small and much extended, having a conspicuous nucleus.

(16) Small and irregularly round. Quite bright and slowly brighter to the middle. Three faint stars following.

NGC	RA	Dec.	nG
128	$00^h26.55'$	$+02^o36'$	8

MCG	RA	Dec.	Mp	NG/IC
0-2-44	$00^h25.85$	$+02^o21'$	15.5	
0-2-45	25.9	$02^o40'$	14.5	
0-2-48	26.1	$02^o34'$	13	NGC.125
0-2-49	26.45	$02^o32'$	14.5	NGC.126
0-2-50	26.5	$02^o36'$	15	NGC.127
0-2-51	26.55	$02^o36'$	12.7	NGC.128
0-2-52	26.5	$02^o36'$	15	NGC.130
0-2-59	27.2	$02^o40'$	14.5	

NGC.128 is a bright edge-on SO galaxy with a peculiar central region which appears to have a four-fold structure on a 5 m plate in the Hubble Atlas of Galaxies. Its close companions NGC.127, and NGC.130 are also shown on the photograph. The group has been observed with 30 inch and 16 inch reflectors, both instruments picking up the 5 NGC objects.

NGC.125
(30) Pretty bright, small and round, having a conspicuous nucleus. Second brightest of 5.

(16) Small and round but quite bright and brighter to the middle. Two stars SP aligned N-S.

NGC.126
(30) Considerably faint, very small and round. Third brightest.

(16) Small and irregularly round. Quite bright and slowly brighter to the middle.

NGC.127
(30) Very faint, very small and round. Close P NGC.128. Faintest of 5.

(16) Almost stellar, very faint. Close P NGC.128.

NGC.128
(30) Considerably bright, pretty large, very much extended edge-on with a conspicuous nucleus. Pointed tips of the lens.

(16) Bright spindle oriented N/S. Much brighter to the middle. Quite a small object being about 1'-1.5' along the major axis.

NGC.130
(30) Very faint, very small and round, possibly having an extremely faint nucleus. Close following NGC.128. Fourth brightest of 5.

(16) Very faint, stellar nebula close following NGC.128.

NGC	RA	Dec.	nG
383	$01^h04.6$	$+32^o08'$	10

MCG	RA	Dec.	Mp	NGC/IC
5-3-48	$01^h04.3$	$32^o30'$	14	NGC.374
5-3-49	04.3	$32^o05'$	15.9	NGC.375
5-3-50	04.45	$32^o14'$	14.0	NGC.379
5-3-51	04.5	$32^o12'$	14.0	NGC.380
5-3-52	04.55	$32^o07'$	15.0	NGC.382
5-3-53	04.6	$32^o08'$	13.6	NGC.383
5-3-55	04.65	$32^o00'$	14.6	NGC.384
5-3-56	04.7	$32^o02'$	14.3	NGC.385
5-3-57	04.7	$32^o05'$	15.7	NGC.386
5-3-59	05.0	$32^o01'$	15.6	NGC.388

The NGC.383 group is a chainlike group mainly composed of elliptical galaxies, and is also known as the Pisces Group. Recent radio observations of the group(1) show that there is an intense radio source, 3C 31, associated with NGC.383 having intense jets aligned along the chain (Arp 331). The NGC.383 group has a redshift similar to the cluster Abell 262(2) and may form part of a supercluster stretching from A262 through the NGC.507 group to the NGC.383 group.

The following galaxies were observed using a 16 inch Newtonian reflector in which the MP field of NGC.383 contains at least six galaxies.

NGC.379 Quite bright and brighter to the middle. Small and round (about 30" in diameter). Close NP NGC.380.

NGC.380 Similar to NGC.379.

NGC.382 Very small and faint. Appears attached to NGC.383 as an obvious protuberance. Star close SP.

NGC.383 Brightest and largest of six. Round, perhaps 45" to 1' in diameter. Quite bright and brighter to the middle with a faint stellar nucleus.

NGC.384 Small and round. Quite bright and brighter to the middle. Close SP NGC.386. Both it and 386 are inferior to NGC.379 + NGC.380.

NGC.386 Similar to NGC.384.

NGC	RA	Dec.	nG
507	$01^h20.95$	$+33^o00$	13

MCG	RA	Dec.	Mp	NGC/IC
5-4-34	$01^h20.2$	32 55	13	NGC.494
5-4-35	20.2	33 13	14.2	NGC.495
5-4-36	20.5	33 17	14	NGC.496
5-4-37	20.5	33 14	16	NGC.498
5-4-38	20.5	33 12	13.2	NGC.499
5-4-39	20.6	33 01	16	IC 1687
5-4-40	20.7	33 05	15	NGC.503
5-4-41	20.7	32 57	14	NGC.504
5-4-44	20.95	33 00	12.8	NGC.507
5-4-45	21.0	33 02	14	NGC.508
5-4-46	21.0	32 48	15	IC 1689
5-4-48	21.25	33 04	16	
5-4-52	21.9	33 13	14	NGC.515

The NGC.507[1] group is a rich condensation in the cloud of galaxies stretching from A262 to the Pisces group. The group was the subject of a paper in WSQJ by Malcom Thompson[2]. His observations are supplemented here by observations made with a 16" Newtonian in UK skies.

NGC.494
($16\frac{1}{2}$) Pretty large, elongated almost E/W. Brightens and widens in the centre to an extended lens. There are three faint stars close south and HP reveals a very faint star on the southern edge just preceding the central area.

(16) Elongated approximately E/W. Quite bright and slowly brighter to the middle. Two stars S and another SF.

NGC.495

(16½) Irregularly round, medium size and brightening in the centre. Resembles NGC.499 in appearance but is smaller and fainter.

(16) Small and round. Quite bright and brighter to the middle. Amongst stars.

NGC.496

(16½) Directly north of NGC.499. Very diffuse with uniform surface brightness. Possible slightly elongated SP/NF. Three very faint stars close preceding.

NGC.499

(16½) The brightest of the group. Irregularly round, large and brightens considerably to a small lens. The outer envelope is diffuse and best seen at HP. A very bright double star lies directly south at a distance, aligned SP/NF.

(16) Bright and much brighter to the middle. Quite large and possibly slightly elongated in PA approximately -45°.

IC 1687

(16½) Well seen at HP. Irregularly round and slightly brighter to the middle. Very close following a faint star.

(16) Very faint and small. Difficult object following a faint star itself following a double star.

NGC.501

(16½) Faint and small, being irregularly round and slightly brighter to the middle.

NGC.503

(16½) Easily visible at HP. Irregularly round with a small brighter central lens.

NGC.504

(16½) Fairly bright, irregularly round. Brightens in the centre to a small nucleus whilst the outer edges are very diffuse.

(16) Small and irregularly round. Quite bright and brighter to the middle.

NGC.507

(16½) Bright, irregularly round and much brighter to the centre, whilst the envelope, which is quite bright, is extensive. The central lens is well developed.

(16) Round and quite large with a diameter of about 1'. Bright and much brighter to the middle. NGC.508 close north.

NGC.508
(16½) Directly north of NGC.507. Quite bright and irregularly round, being much brighter to the centre with quite extensive envelope. Star close NF.

(16) Close N of NGC.508. Quite bright, round and much brighter to the middle. Smaller and fainter than NGC.507. Star close north.

IC 1689
(16½) Close south following NGC.507. An extremely faint, small nebulous knot of uniform surface brightness.

MCG 5-4-45
(16½) North following NGC.507. Preceded by two stars. Of medium size and irregular in shape, brightening to the centre. Well seen at HP.

WS	NGC	RA	Dec.	nG
	529	$01^h22.9'$	$+34^o27'$	5

MCG	RA	Dec.	Mp	NGC/IC
6-4-19	$01^h22.65'$	$34^o27'$	13.1	NGC.529
6-5-20	23.6'	$34^o29'$	15	
6-4-21	23.65'	$34^o26'$	13.2	NGC.536
6-4-22	23.75'	$34^o24'$	15	NGC.542
not in MCG	23.25'	$34^o28'$	15.0	RNGC 531

NGC.528

(16½) First of four. Small, round and much brighter to the middle, being quite bright overall. At a distance south following is a very bright star.

NGC.531
(16½) Very faint, very small. A nebulous knot with a star close south. Closer to 536 than 529.

NGC.536
(16½) The largest of the group and brightens in the middle to a small lens surrounded by an irregular halo ending abruptly on the northern edge.

NGC.542
(16½) Very close south following NGC.536, being extremely faint and narrow elongated NP/SF. Seen with difficulty.

NGC	RA	Dec.	nG
978	$02^h31.1$	$+32^o45'$	5

MCG	RA	Dec.	Mp	NGC/IC
5-7-8	31.1	32 45	13.5	NGC.969
5-7-9	31.1	32 47	15.5	NGC.970
5-7-12	31.3	32 45	13.9	NGC.974
5-7-16	31.8	32 39	13.3	NGC.978(a)
5-7-17	31.8	32 39	15	NGC.978b

NGC.969
(16½) Very small, irregularly round and a little brighter to the middle. A fairly bright star lies north following at a distance.

(16) Quite bright and much brighter to the middle. Round, diameter about 30". Faint stars south.

NGC.970
(16½) Very small and irregularly round. Faint and brighter in the middle.

(16) Very faint nebula slightly elongated N/S. Faint star involved north.

NGC.974
(16½) Largest in the group being elongated SP/NF. Close following are two stars.

(16) Quite bright and slightly elongated E/W. Star mag. 10 following.

NGC.978
(16½) The brightest in the group. Irregularly round in shape and appearing brighter on the preceding side. A faint star of nebulous knot follows very close south. (MCG 5-7-17).

NGC	RA	Dec.	nG
4005	$11^h55.5'$	$+25^{\circ}23'$	

MCG	RA	Dec.	Mp	NCG/IC	Type
4-28-99	$11^h54.7$	$25^{\circ}28'$	14.4	NGC.3987	Sb
4-28-100	54.8	$25^{\circ}30'$	15.3	NGC.3989	
4-28-101	55.0	$25^{\circ}31'$	14.8	NGC.3993	S
4-28-102	55.2	$25^{\circ}32'$	14.3	NGC.3997	Sb
4-28-103	55.3	$25^{\circ}25'$	15.2	NGC.4000	
4-28-107	55.5	$25^{\circ}23'$	14.1	NGC.4005	S
4-28-108	56.0	$25^{\circ}35'$	14.7	NGC.4018	
4-28-109	56.0	$25^{\circ}19'$	} 14.2	} NGC.4015	Double
4-28-110	56.0	$25^{\circ}19'$			
4-28-111	56.3	$25^{\circ}30'$	14.4	NGC.4022	
4-28-112	56.35	$25^{\circ}21'$	15.3	NGC.4021	
4-28-113	56.4	$25^{\circ}16'$	14.6	NGC.4023	

The NGC.4005 group, together with the groups associated with NGC.4065 and NGC.4092, have been shown to be probably condensations in the supercluster stretching from A1656 to A1367 across about 11° of the sky. All three groups are rather rich and present some of the most spectacular galaxy fields available to moderate apertures.

NGC.3987

(30) Considerably bright and much extended, having a conspicuous nuclear bulge. Two way faint stars close east. Brightest of 4 in the north west quadrant of the group.

($16\frac{1}{2}$) First of 6 galaxies. Quite large and greatly extended in PA 58°, being very narrow except at the centre where it both widens and brightens.

(16) Spindle about 1.5' × 20" with a PA of about -45°. Quite bright and much brighter to the middle. Star of mag. 9 or 10 3' N.

NGC.3989
(30) Considerably faint, small and round. Diffuse texture. Faintest of NW subgroup.

NGC.3993
(30) Pretty bright, pretty large and much extended, having a diffuse texture. Third brightest of NW subgroup.

(16½) North following NGC.3987. Elongated in PA 141°. Very narrow, and of medium size. Brightens towards the centre. Amongst stars.

(16) Small and irregularly round. Quite bright and brighter to the middle. Two stars to the north.

NGC.3997
(30) Pretty bright, pretty large, very much extended and mottled with a faint nucleus. Pretty bright star close east. Second brightest of NW subgroup.

(16½) North following NGC.3993. Fairly bright, irregular in shape with its major axis east/west. Brightens to the middle. Between two stars.

(16) Small and slightly elongated east/west. Quite bright and considerably brighter to the middle. Between two stars.

NGC.3999
(30) Faint, very small and round, possible having a faint stellar nucleus.

NGC.4000
(30) Considerably faint, small and very much extended N/S.

NGC.4005
(30) Considerably bright, small and round, possibly with a faint nucleus.

(16½) South following 3987, 3993 and 3997 is a bright star and 4005 lies close south following this star. It is irregularly round and of uniform surface brighteness except that it brightens slowly to the centre.

(16) Bright and much brighter to the middle. Bright star north preceding.

NGC.4011
(30) Faint, very small and round.

NGC.4015

(30) Pretty bright, small and round having a stellar nucleus. Brightest of three.

(16½) Fairly large and irregularly round. Quite bright and easily seen. Brighter to the middle.

(16) Quite bright and brighter to the middle. Perhaps slightly elongated in PA 45°.

NGC.4018

(30) Pretty bright, very much extended and having a mottled texture.

NGC.4021

(30) Considerably faint, small and round having a nucleus. Faintest of three.

(16) North following 4015. Small and round, quite bright and brighter to the middle.

NGC.4022

(30) Pretty bright, small and round. Conspicuous nucleus with a faint halo.

(16) Small and round. Quite bright and brighter to the middle.

NGC.4023

(30) Pretty bright, nearly round and small. Diffuse with a very faint star nearly due north.

(16½) Very faint and of medium size, being slightly brighter to the middle.

(16) South of NGC.4015. Irregularly round and larger and brighter than 4021. Some extremely faint objects near by add much visual interest. (RNGC mentions two stars close north.) Quite bright and brighter to the middle.

NGC	RA	Dec.	nG
5353	$13^h51.3$	$+40^o32'$	5

MCG	RA	Dec.	Mp	NGC/IC	Type
7-29-9	$13^h51.2$	$40^o36'$	12.4	NGC.5350	SBb-c
7-29-10	51.3	$40^o32'$	11.8	NGC.5353	SO
7-29-11	51.3	$40^o33'$	12.3	NGC.5354	SO
7-29-12	51.6	$40^o35'$	14.0	NGC.5355	SO?
7-29-13	51.8	$40^o32'$	14	NGC.5358	

NGC. 5350
($16\frac{1}{2}$) First of four galaxies and the most northerly. Fairly large irregularly round and brightening in the middle. A bright star precedes.

(16) Bright and quite large, slowly brightening to the middle. Oval elongated in PA of about 50°, although different observations yield different orientations. Beautiful double star preceding, the brighter being a brilliant yellow whilst the SP fainter component is electric blue. Easy in a 10 inch.

NGC.5353
($16\frac{1}{2}$) Brightest of the group. Slightly elongated NP/SF. Stellar nucleus.

(16) Bright and suddenly much brighter to the middle to a stellar nucleus. Spindle about 1' × 30" elongated approximately N-S. NGC.5354 very close north.

NGC.5354
($16\frac{1}{2}$) Small, round and suddenly brighter to the centre. Third of four.

(16) Bright but inferior to NGC.5353, otherwise almost a carbon copy.

NGC.5355
($16\frac{1}{2}$) Fourth and faintest of four. Round, small and brightening slightly to the centre.

(16) Quite bright, small and slightly elongated approximately E-W. Visible in a 10 inch with some effort.

NGC	RA	Dec.	nG(1°)
5416	$13^h59.7$	+09°41'	

The NGC.5416 group has been of some professional interest recently[1] because of its possible membership of the Coma - A1367 supercluster. The following data is based upon reference (1), and refers to galaxies with a 1° diameter circle roughly centred upon NGC.5423. The group just fails to meet the population criterion for inclusion in Abell's richness group 0.

MCG	RA	Dec.	Mp	PA	Type	NGC/IC
2-36-6	$13^h57.7$	09°13'	14.8	121°	Sbc	
	57.8	09°33'	15.6	142°	Sc	
	58.0	08°55'	15.6	065	Sc	
	58.0	09°09'	15.6	104	E	
	58.7	08°51'	15.3	135	S/SO	
	58.8	10°22'	15.3	130	Sc	
	59.0	09°25	15.7	157	?	
	59.2	10°07'	15.2	006	SO/a	
2-36-9	59.3	09°44'	14.4	132	Scd	NGC.5409
2-36-10	59.4	09°02'	14.7	-	Sb	
2-36-11	59.5	09°11'	14.7	035	?	NGC.5411
	59.5	09°47'	15.6	118	S	
2-36-13	59.6	10°11	13.9	079	SOp	NGC.5414
	59.7	09°18'	15.7	003	Ir	
2-36-14	59.7	09°41'	13.6	011	Sc	NGC.5416
2-36-15	$14^h00.1$	09°19'	15.0	121	Sd	
	00.1	09°22'	15.6	179	SC	
	00.2	09°25'	15.7	159	Sd	
2-36-16	00.2	09°36'	15.5	141	?	
2-36-17	00.3	09°35'	13.9	058	E	NGC.5423
	00.4	09°02'	15.6	024	Scd	
	00.4	09°09'	15.6	007	Sc	

MCG	RA	Dec.	Mp	PA	Type	NGC/IC
2-36-18	$13^h00.4$	09°36'	15.5	174	SO	
2-36-19	00.5	09°40'	14.3	014	SOp	NGC.5424
	00.6	09°12'	15.1	-	E/SO	
2-36-20	00.6	09°37'	14.8	130	Sd	NGC.5431
2-36-21	00.7	09°09'	15.0	065	SO	
2-36-22	00.9	09°41'	14.3	-	Sc	RNGC 5434A
2-36-23	01.0	09°28'	14.8	070	?	
2-36-24	01.0	09°43'	14.7	157	SC	RNGC 5434B
2-36-25	01.2	09°49'	14.9	027	SO	NGC.5436
2-36-28	01.3	09°45'	15.1	086	S/SO	NGC.5437
2-36-29	01.3	09°51'	14.7	171	E	NGC.5438
2-36-31	02.3	09°04'	14.9	093	Sa	
	02.6	09°35'	15.6	142	Sm	
	02.8	09°52'	15.5	039	SO/a	
	03.1	09°09'	15.3	-	Ir	
	03.3	09°45'	15.7	054	S	
	03.5	09°16'	15.2	095	SO/a	
2-36-40	03.7	09°36'	14.5	133	SO	NGC.5463
	03.7	09°36'	15.4	026	E	
2-36-41	04.5	10°42'	15.4	052	SOp	

NGC.5409

(16) First of a MP field of two. Small oval about 45" along major axis, oriented approximately N-S. Quite bright and slowly brighter to the middle.

NGC.5416

(16) Round, about 45" in diameter. Quite bright and brighter to the middle. Second of a field of two. Brightest and largest in group.

NGC.5423
(16) First of a MP field of two. Quite bright and slowly brighter to the middle, but fainter than NGC.5424. It is also smaller than 5424. Round, about 25" in diameter.

NGC.5424
(16) Second of two, also first of three including NGC.5434. Round, about 30" in diameter. Quite bright and slowly brighter to the middle. Faint star very close to the south.

RNGC 5434A
(16) Round, about 45" in diameter. Quite bright and of almost uniform low surface brightness, brightening very slowly to the middle. RNGC 5434B close north following.

RNGC 5434B
(16) Third of three. Faint with low surface brightness. Cigar close north following RNGC 5434A elongated approximately E-W. About 45" ×20".

NGC.5436
(16) First of three. Round and quite small (about 30"). Quite bright and brighter to the middle.

NGC.5437
(16) Second of three and faintest in the field. Faint and of low surface brightness. Very small and round.

NGC.5438
(16) Third of three. Very similar to NGC.5436 preceding by 4 or 5'.

NGC	RA	Dec.	nG
7230	$22^h33.65'$	$+33^o39'$	5

The NGC.7230 group is otherwise known as 'Stephan's Quintet', and has formed the topic of innumerable research papers concerned with the membership or non-membership of NGC.7320 in the group. Perhaps the best reference to this controversial group is 'The Redshift Controversy' edited by Field, Arp and Bahcall.

The group is a favourite target for amateur astronomers, and observations have been claimed by observers using telescopes from 5 inch aperture upwards. The group is difficult visually, and its field is rich in stars. One would expect that the individual galaxies would need at least an 8 inch in good skies for any success. The following observations were made by observers using 8½, 12 and 16 inch Newtonian reflectors.

MCG	RA	Dec.	Mp	NGC/IC
6-49-38	$22^h33.5'$	$33^o39'$	15.3	NGC.7317
6-49-39	33.6'	$33^o40'$	14.8	NGC.7318a
6-49-40	33.6'	$33^o40'$	14.9	NGC.7318b
6-49-41	33.65'	$33^o41'$	13.7	NGC.7319
6-49-42	33.65'	$33^o39'$	13.6	NGC.7320
6-49-43	33.9'	$33^o42'$	17	RNGC 7320C
6-49-44	33.9'	$34^o15'$	15	
	34.15'	$33^o29'$	?	RNGC 7320A
	35.05'	$33^o37'$	?	RNGC 7320B

NGC.7317

(16) Very faint, almost stellar with a star in contact. Difficult.

(12½) Small, round nebulosity perhaps equal in brightness to NGC.7319. Difficult without averted vision because of its proximity to a 12th magnitude star.

(8½) A small, faint glow 1' SW of NGC.7320.

NGC.7319a/b
(16) Small elongated glow oriented roughly EW embedded in which are two almost stellar nuclei. Very close NP NGC.7320.

(12½) Brightest object in the group. Fairly well condensed nebula with two faint stellar nuclei aligned in PA 45-225°, the following being brighter.

(8½) Not found.

NGC.7319
(16) Irregularly round nebula to the north of NGC.7320. Quite bright and a little brighter to the middle.

(12½) A faint circular nebula, small and without any sign of condensation.

(8½) Not found.

NGC.7320
(16) Easily the brightest of the group. Oval elongated roughly EW. Quite bright overall with a diffuse texture brightening slightly to the middle. Faint stellar nucleus intermittantly visible.

(12½) Oval nebula elongated in PA 135°-315°. Much larger than the other members and most easily visible after NGC.7318a/b. It shows little condensation but there is a faint, centrally located stellar nucleus.

(8½) Faint streaky glow elongated in PA 130°-310°.

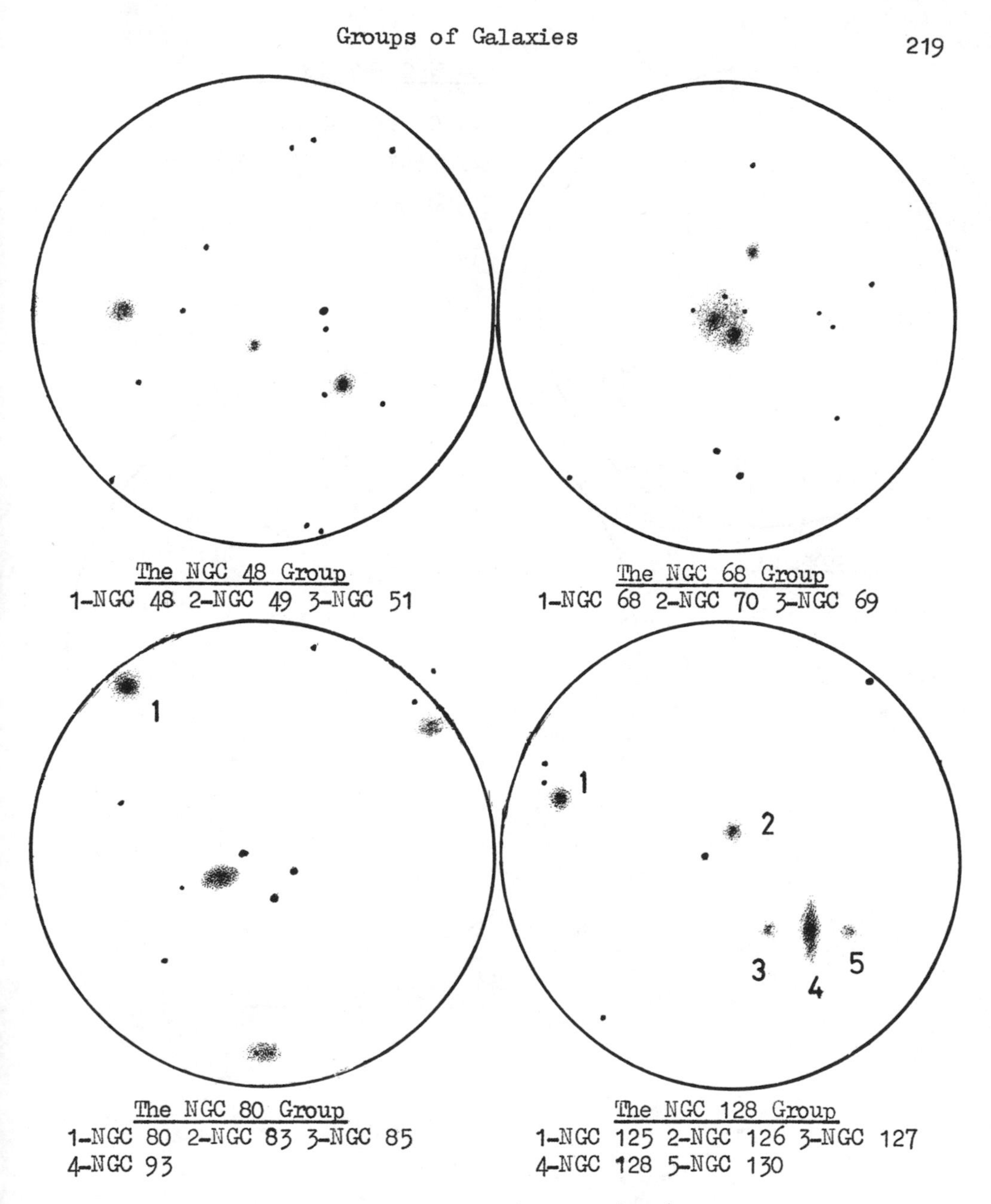

The NGC 48 Group
1-NGC 48 2-NGC 49 3-NGC 51

The NGC 68 Group
1-NGC 68 2-NGC 70 3-NGC 69

The NGC 80 Group
1-NGC 80 2-NGC 83 3-NGC 85
4-NGC 93

The NGC 128 Group
1-NGC 125 2-NGC 126 3-NGC 127
4-NGC 128 5-NGC 130

GSW 16-inch f/5 x160 (12')

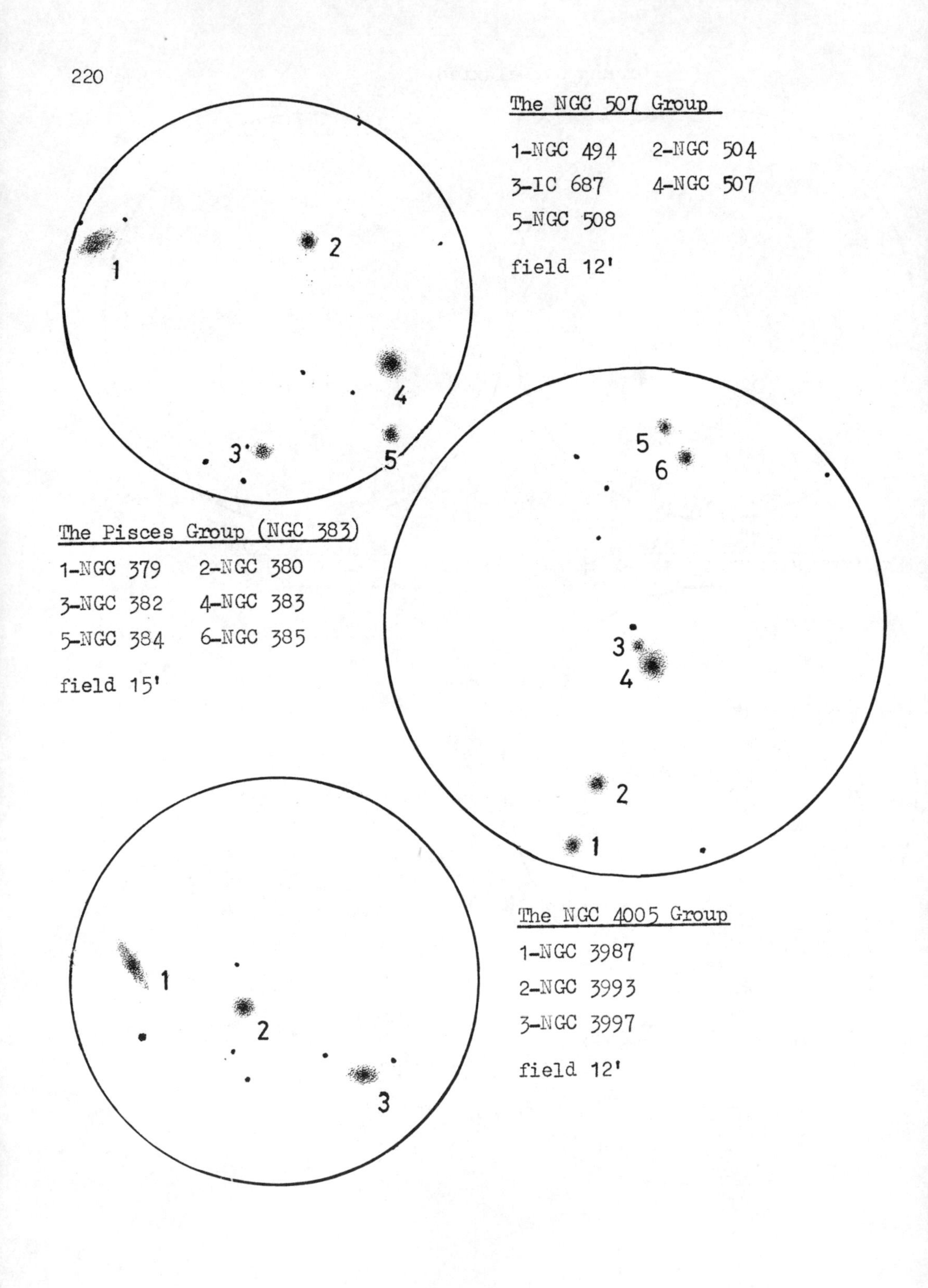
The NGC 507 Group
1-NGC 494
2-NGC 504
3-IC 687
4-NGC 507
5-NGC 508
field 12'
1
2
3
4
5
The Pisces Group (NGC 383)
1-NGC 379
2-NGC 380
3-NGC 382
4-NGC 383
5-NGC 384
6-NGC 385
field 15'
1
2
3
4
5
6
The NGC 4005 Group
1-NGC 3987
2-NGC 3993
3-NGC 3997
field 12'
1
2
3

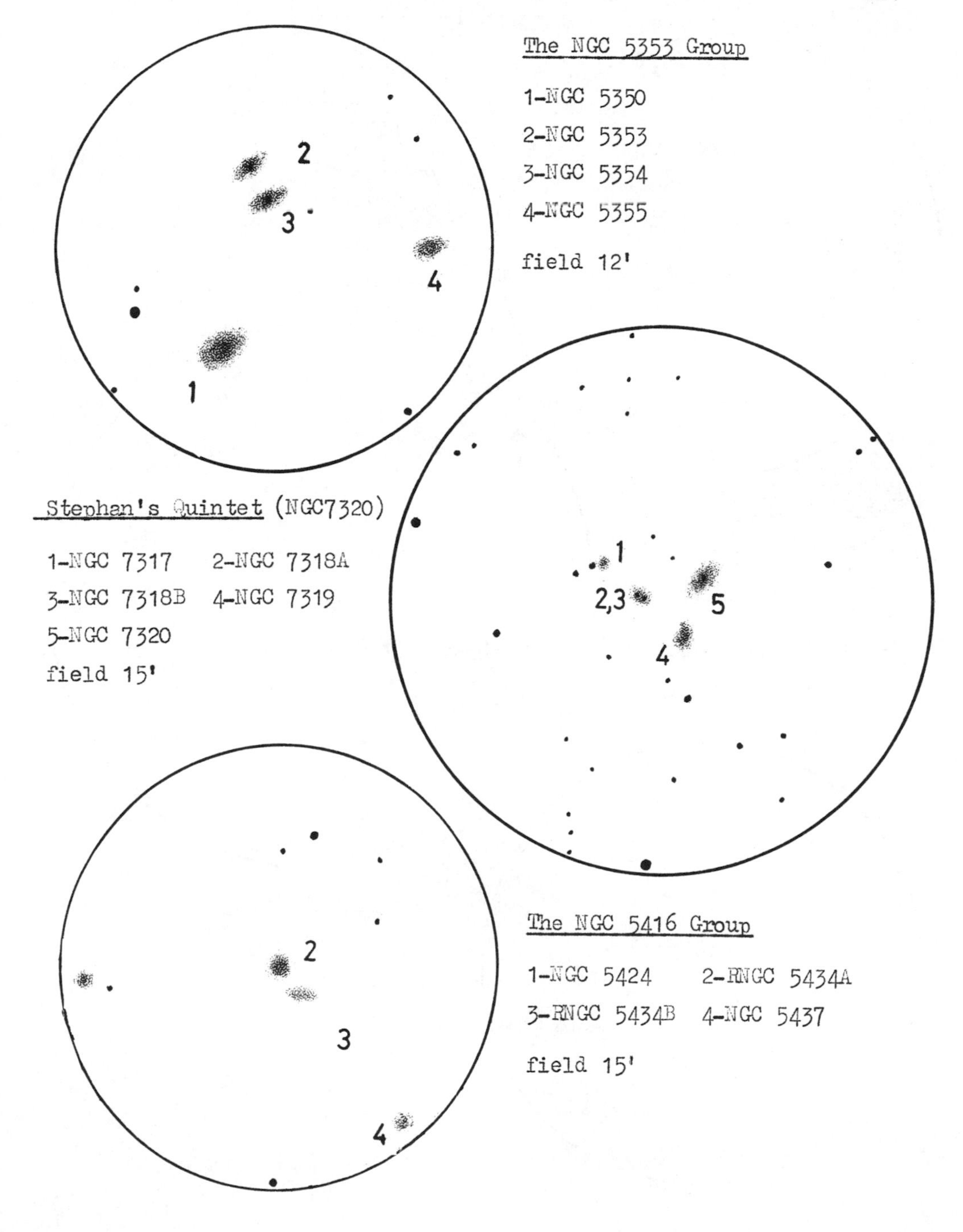
The NGC 5353 Group
1-NGC 5350
2-NGC 5353
3-NGC 5354
4-NGC 5355
field 12'
1
2
3
4
Stephan's Quintet (NGC7320)
1-NGC 7317
2-NGC 7318A
3-NGC 7318B
4-NGC 7319
5-NGC 7320
field 15'
1
2,3
4
5
The NGC 5416 Group
1-NGC 5424
2-RNGC 5434A
3-RNGC 5434B
4-NGC 5437
field 15'
2
3
4

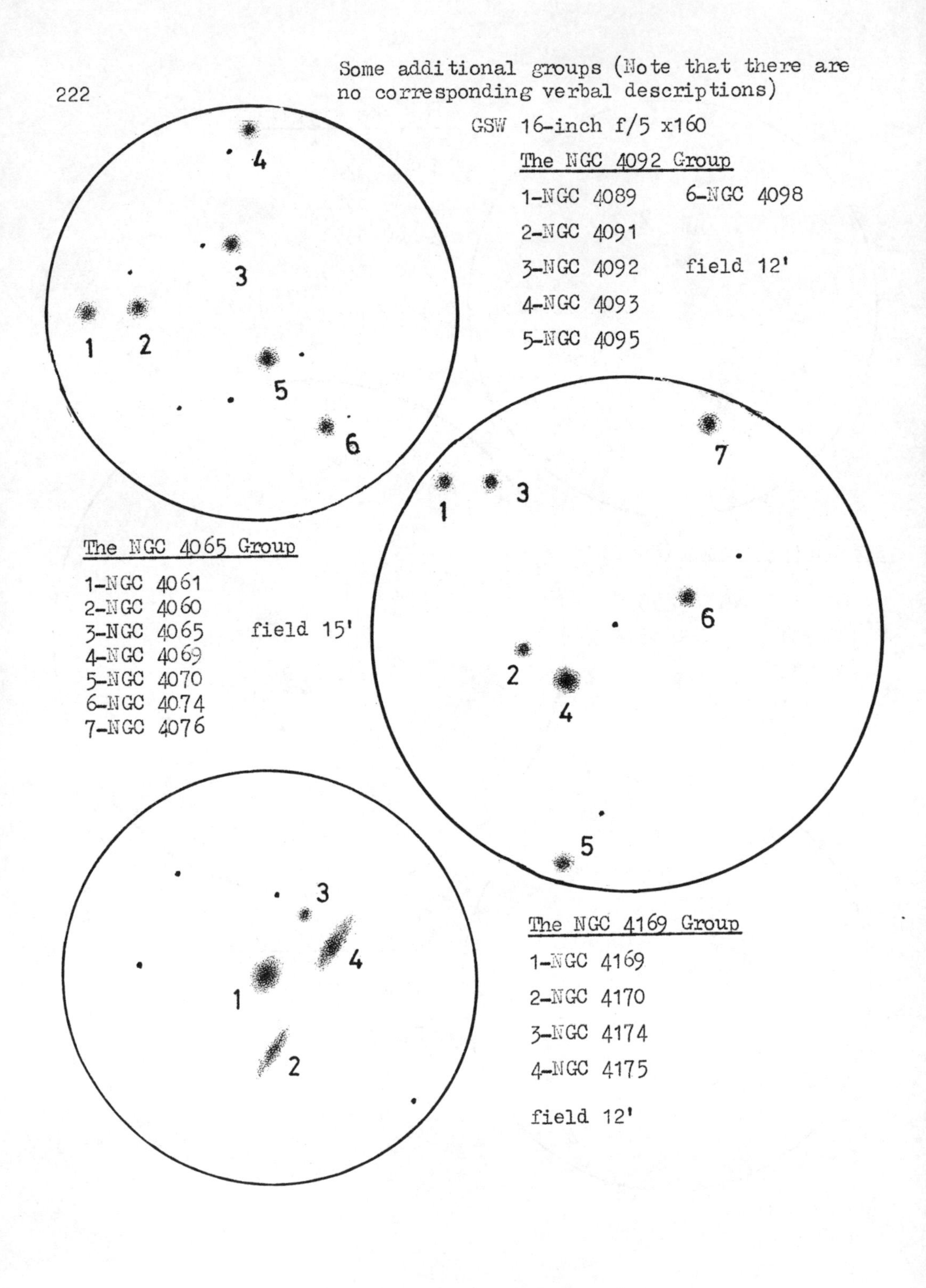
Some additional groups (Note that there are no corresponding verbal descriptions)
GSW 16-inch f/5 x160
The NGC 4092 Group
1-NGC 4089
2-NGC 4091
3-NGC 4092
4-NGC 4093
5-NGC 4095
6-NGC 4098
field 12'
The NGC 4065 Group
1-NGC 4061
2-NGC 4060
3-NGC 4065
4-NGC 4069
5-NGC 4070
6-NGC 4074
7-NGC 4076
field 15'
The NGC 4169 Group
1-NGC 4169
2-NGC 4170
3-NGC 4174
4-NGC 4175
field 12'
1
2
3
4
5
6
7

Large Telescope Observations

The following observations were made by R.J. Buta, a staff astronomer of the McDonald Observatory of the University of Texas, using 30 and 36 inch telescopes visually.

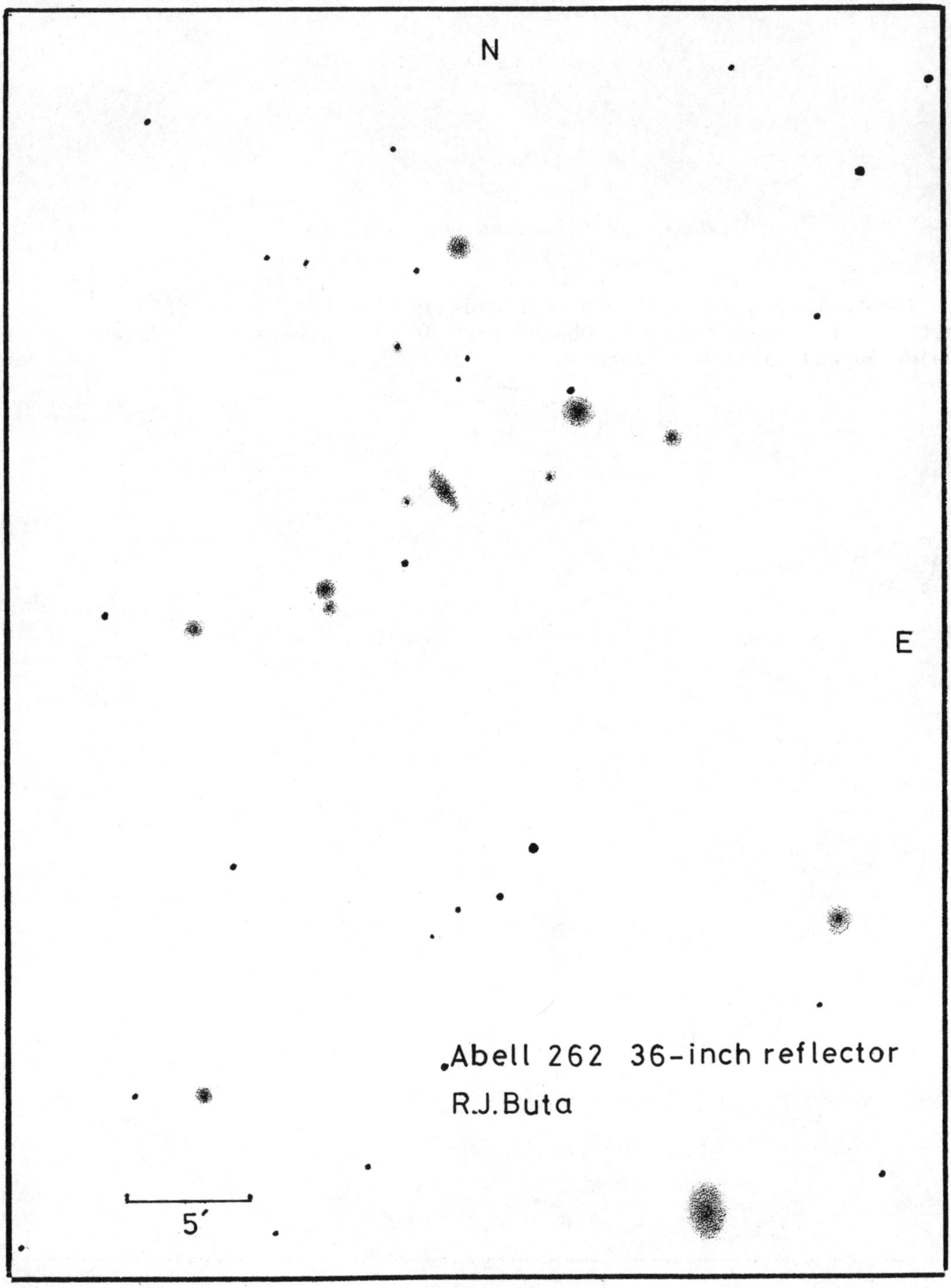

N

E

Abell 262 36-inch reflector

R.J. Buta

5′

N

E

The Perseus Cluster Abell 426

36-inch reflector. R.J. Buta.

5′

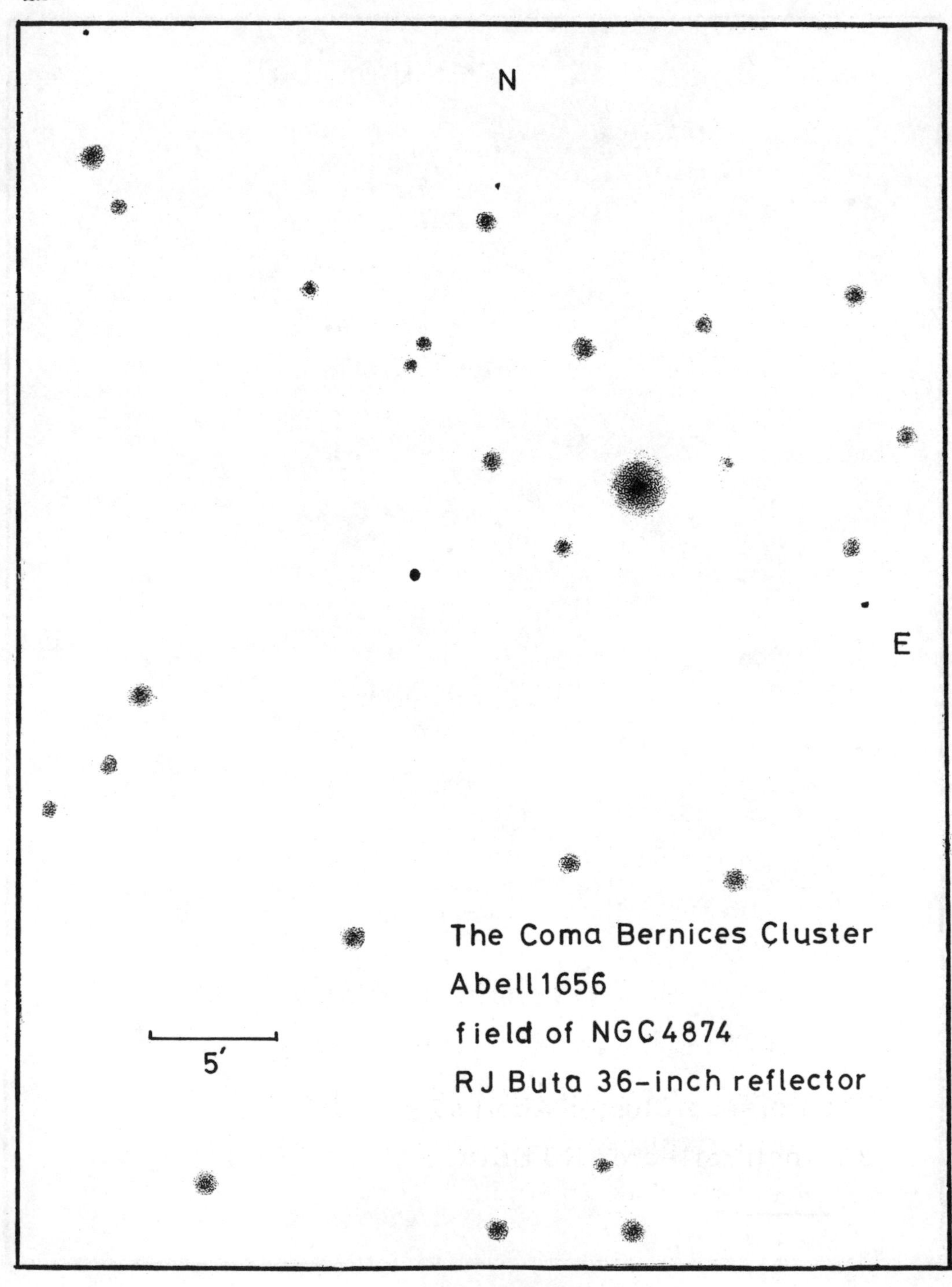
N
E
5′
The Coma Bernices Cluster
Abell 1656
field of NGC 4874
RJ Buta 36-inch reflector

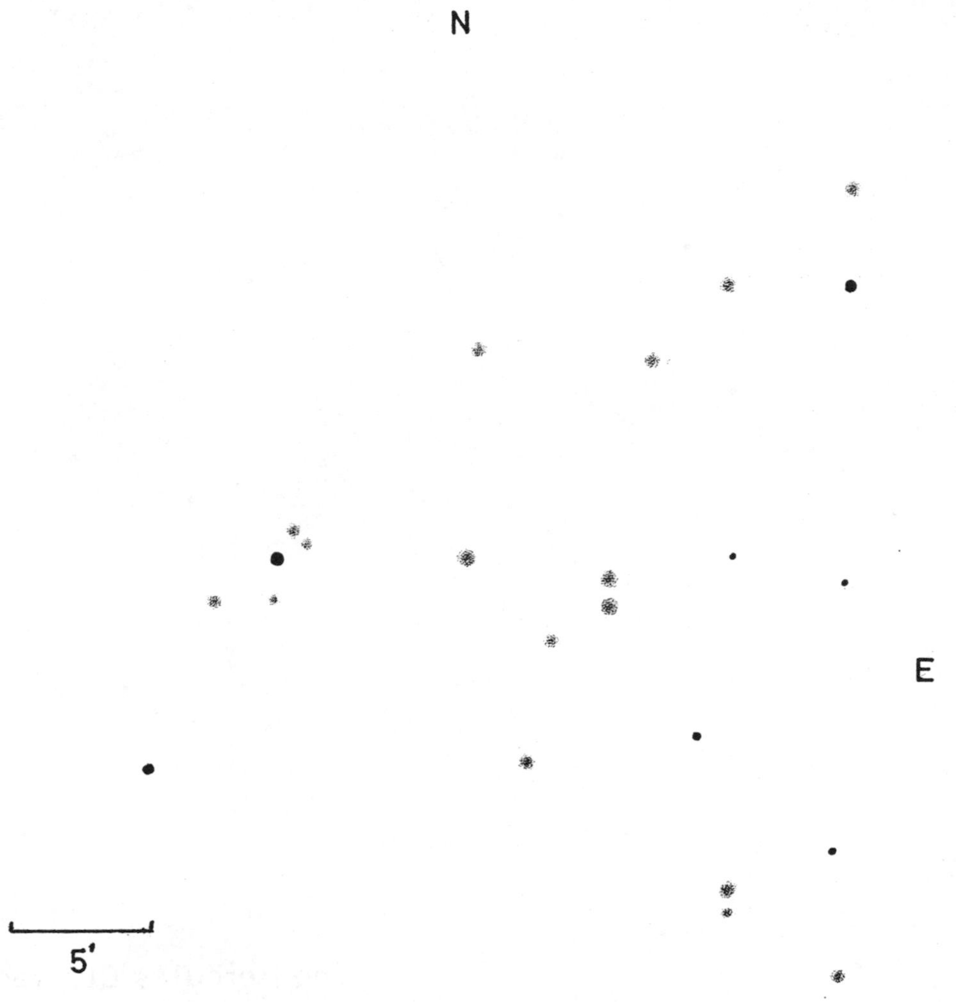

N

E

5'

The Corona Borealis Cluster Abell 2065
R.J.Buta 36-inch reflector.
Seen through a thin haze. Cluster of eef
ees galaxies in Corona Borealis. Not
easy in the 36-inch.

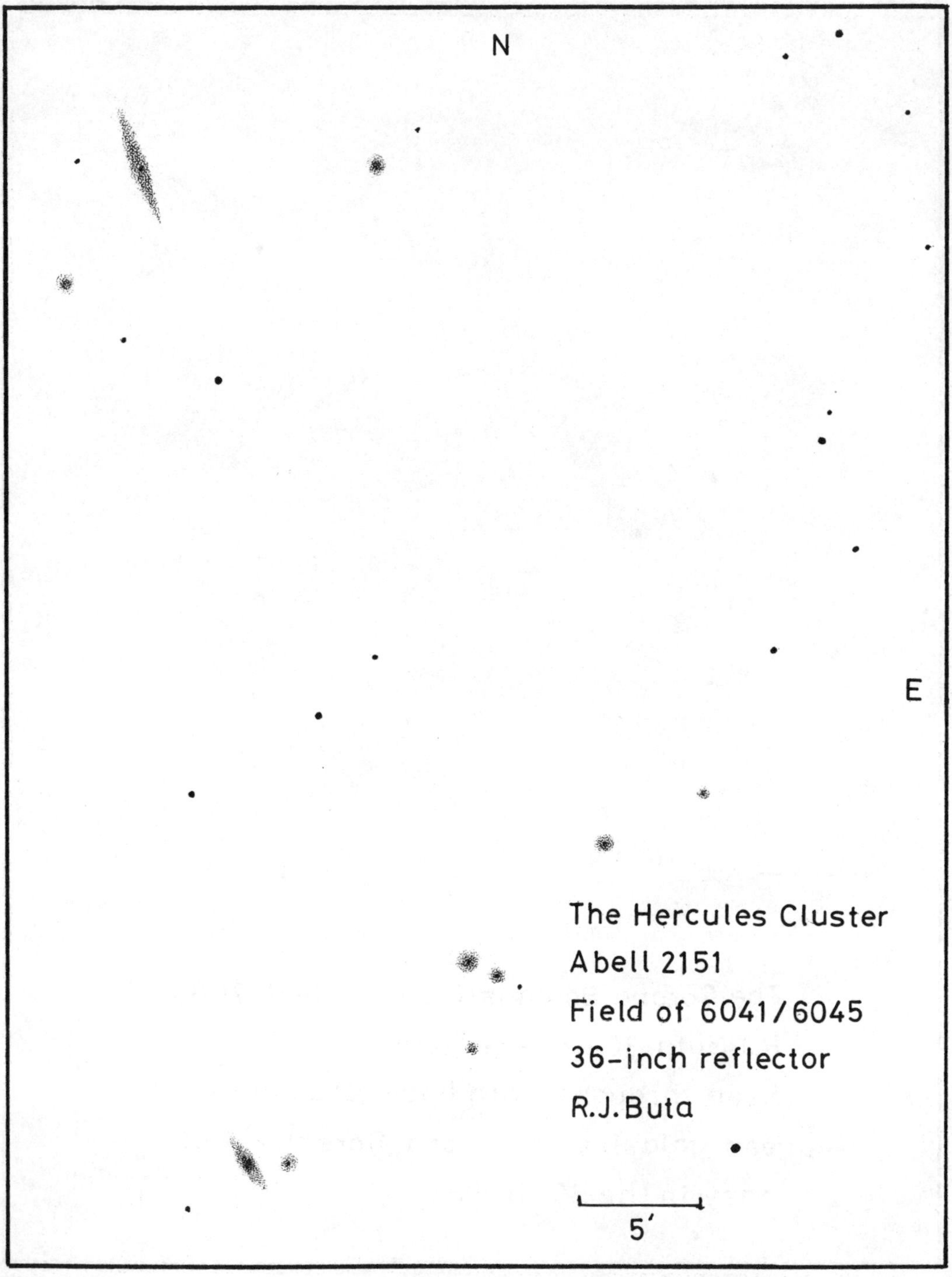

N

E

The Hercules Cluster
Abell 2151
Field of 6041/6045
36-inch reflector
R.J. Buta

5′

The Hercules Cluster
Abell 2151
Field of NGC 6050
36-inch reflector
R J Buta

N

E

5′

N

E

Abell 2199
Field of NGC 6166
30-inch reflector
R.J. Buta

5′

N

E

The NGC 80 Group
30-inch reflector
R.J. Buta

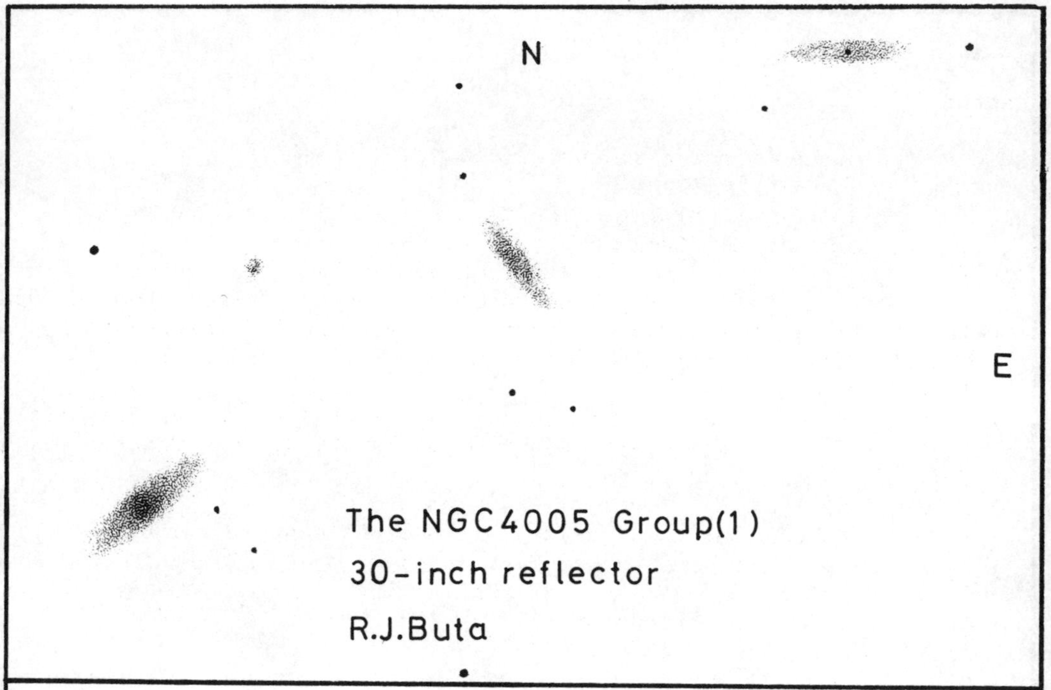

N

E

The NGC4005 Group(1)
30-inch reflector
R.J.Buta

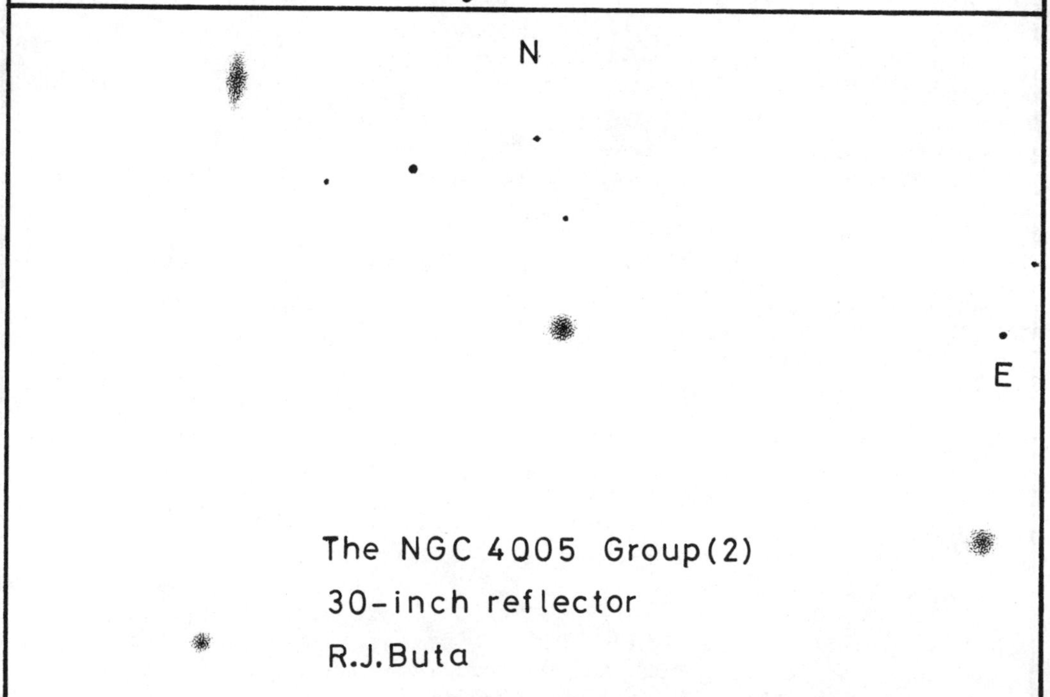

N

E

The NGC 4005 Group(2)
30-inch reflector
R.J.Buta

Part (1)

Chapter (2)

(1) K. Glyn Jones 'Messier's Nebulae and Star Clusters'
Faber and Faber, London 1968
(2) J.H. Mallas and E. Kreimer 'The Messier Album'
Sky Publishing Co., Cambridge Mass. 1978
(3) J.L.E. Dreyer 'New General Catalogue of Nebulae and Clusters of Stars' Memoirs of the Royal Astronomical Society. RAS. London 1962
(4) J.W. Sulentic, W.G. Tifft 'The Revised New General Catalogue of Non-Stellar Astronomical Objects' University of Arizona Press, Tucson 1973
(5) H. Shapley and A. Ames 'A Survey of the External Galaxies Brighter than the Thirteenth Magnitude' Annals of Havard College Observatory, Vol. 88, No. 2, 1932
(6) F. Zwicky et al. Catalogue of Galaxies and Clusters of Galaxies' (6 vols.) Pasadena, California Institute of Technology 1961-1968
(7) B.A. Vorontsov-Velyaminov and V.P. Arhipova 'Morphological Catalogue of Galaxies' (4 vols.) University of Moscow (1964)
(8) G.O. Abell 'The Distribution of Rich Clusters of Galaxies'
Ap. J. Suppl. No. 3, pp. 211-288 (1958)
(9) G.B. Field, H.C. Arp and J.N. Bachcall 'The Redshift Controversy'
W.A. Benjamin Inc., Reading Mass. 1973
(10) H.J. Rood 'Nearby Groups of Galaxy Clusters'
Ap. J. No. 207, pp. 16-24, 1976
(11) P.J.E. Peebles et al. 'The Clustering of Galaxies'
Scientific American November 1977
(12) S.A. Gregory and L.A. Thompson 'The Coma/A1367 Supercluster and its Environs'
Ap. J. No. 222, pp. 784-799 (1978)
(13) G. Chincarini and H.J. Rood 'Space Distribution of Sc Galaxies: Clues to the Large Scale Structure'
Ap. J. No. 230, pp. 648-654 (1979)

Chapter (3)

(1) J.E. Gunn and J.R. Gott III 'On the Infall of Matter into Clusters of Galaxies and Some Effects on Their Evolution'
Ap. J. No. 176, pp. 1-19 (1972)
(2) R.J. Mitchell, J.C. Ives and J.L. Culhane 'The X-Ray Temperature of Eight Clusters of Galaxies and Their Relationship to Other Cluster Properties'
Mon. Not. R.A.S., No. 181, pp. 26-35 (Short Communications) (1977)
(3) G.R. Gisler and G.K. Miley '610 MHz Observations of the Perseus Cluster of Galaxies with the Westerbork Synthesis Radio Telescope'
Astrom. Astrophys. No. 76, pp. 109-119 (1979)
(4) F.N. Owen and L. Rudnick 'Radio Sources with Wide Angle Tails in Abell Clusters of Galaxies'
Ap. J. No. 205, pp. L1-L4, (1976)

(5) E.A. Valentin et al., 'A Westerbork Survey of Clusters of Galaxies IV' Observations of the Coma Cluster at 610 MHz!
Astron. Astrophys. Suppl. No. 28, pp. 333-349 (1977)
(6) R.J. Wellington et al. 'High Resolution Map of NGC.1265'
Nature Vol. 244, August 24, (1973)
(7) J.S. Gallagher 'Possible Optical Evidence for Ram-Pressure Sweeping in the Hydra-I Cluster of Galaxies'
Ap. J. No. 223, pp. 386-390 (1978)
(8) P.W. Sanford and J.C. Ives 'Ariel Results on Extragalactic X-Ray Sources' Proc. Roy. Soc. A, No. 350, pp. 491-503 (1976)
(9) K.M. Strom and S.E. Strom 'Surface Brightness and Colour Distributions of E and SO Galaxies I - The Coma Cluster II - A 426 and A 1367'
Astron. J. Vol. 83, No. 2, pp. 73-134 (1978)
Astron. J. Vol. 83, No. 7, pp. 732-763 (1978)
(10) H. Butcher and A. Oemler Jr. 'The Evolution of Galaxies in Clusters I. ISIT Photometry of C1.0024 + 1654 and 3C295'
Ap. J. No. 219, pp. 18-30 (1978)
(11) A. Dressler 'A Comprehensive Study of 12 Very Rich Clusters of Galaxies I and II'
Ap. J. No. 223, pp. 765-787 (1978)
Ap. J. No. 226, pp. 55-69 (1978)
(12) A. Oemler Ap. J. No. 180, p. 11 (1973)
(13) D. Carter 'The Optical Extent of Giant E and cD Galaxies'
Man. Not. R.A.S., No. 178, pp. 137-148 (1977)
(14) L.A. Thompson 'The Angular Momentum Properties of Galaxies in Rich Clusters'
Ap. J. No. 209, pp. 22-34 (1976)
(15) J.G. Godwin et al. 'Studies of Rich Clusters of Galaxies V'
University of Oxford Preprint 1979 (to appear in Mon. Not. R.A.S.)
(16) D. Trevese and A. Vignato 'A Dynamical Condition for a Relativistic Galaxy Cluster Model'
Astrophysics and Space Science 41, pp. 213-219 (1976)

Part (2)

(1) The Virgo Cluster

(1) K. Glyn Jones ' Messier's Nebulae and Star Clusters'
Faber and Faber, London 1968
(2) L.S. Copeland 'Adventuring in the Virgo Cloud'
Sky and Telescope February 1955
(3) J. Newton 'Deep Sky Objects A Photographic Guide for the Amateur'
Gall Publications Toronto (1977)
(4) J.H. Mallas and E. Kreimer 'The Messier Album'
Sky Publishing Co., Cambridge Mass. 1978

(2) Abell Clusters

Abell 119

(1) D. Maccagni et al. 'An X-Ray and Optical Study of 7 Clusters of Galaxies' Astron. Astrophys., 6, pp. 127-133 (1978)

Abell 194

(1) G. Chincarini and H.J. Rood 'The Structure of the Galaxy Cluster A 194' Ap. J. No. 214, pp. 351-358 (1977)
(2) S.M. Simkin 'Optical Properties of the Radio Source PKS 0123-01 (3C40) in Abell 194'
(3) H. Shapley 'Galaxies' 3rd ed. (Revised by P.W. Hodge) Harvard University Press (1972)

Abell 262

(1) C. Moss and R.J. Dickens 'Redshifts of Galaxies in the Cluster A262 and in the Region of the Pisces Group (NGC.383)' Mon. Not. R.A.S. 178, pp. 701-715 (1977)

Abell 347

(1) P. Hintzen and W.R. Oegerle 'Observations of Clusters Containing Radio Tail Galaxies' Astron. J. Vol. 83, No. J, pp. 478-481 (1978)

Abell 426

(1) G. Chincarini and H.J. Rood 'Dynamics of the Perseus Cluster of Galaxies' Ap. J. No. 168, pp. 321-325 (1971)
(2) P. Veron 'NGC.1275' A BL Lacerta Object?' Nature Vol. 272, p. 430 (1976)
(3) V.C. Rubin et al. 'New Observations of the NGC.1275 Phenomenon' Ap. J. No. 211, pp. 693-696 (1977)
(4) V.C. Rubin et al. 'A New Mapping of the Velocity Field of NGC.1275' Ap. J. Suppl 37, pp. 235-249 (1978)
(5) W. Cash and R.F. Malina 'A Soft X-Ray Map of the Perseus Cluster of Galaxies' Ap. J. Lett. 209, pp. L111-L114 (1976)
(6) G.R. Gisler and G.K. Miley '610 MHz Observations of the Perseus Cluster of Galaxies with the Westerbork Synthesis Radio Telescope' Astron. Astrophys. No. 76, pp. 109-119 (1979)
(7) I.I.K. Pauliny-Toth et al. 'High Resolution Observations of NGC.1275 with a Four-Element Intercontinental Radio Interferometer' Nature Vol. 259 January 1 and 8 (1976)

Abell 1185

(1) H.C. Arp 'Observational Paradoxes in Extragalactic Astronomy' Science 17 December 1971, Vol. 174, No. 4015, pp. 1189-1200

Abell 1228

(1) K.M. Strom and S.E. Strom 'Surface Brightness and Colour Distributions of E and SO Galaxies III' Astron. J. Vol. 83, No. 11, pp. 1293-1330 (1978)

Abell 1367

(1) R.J. Dickens and C. Moss 'Redshifts of Galaxies in the Cluster A 1367' Mon. Not. R.A.S. 174, pp. 47-58 (1976)
(2) S.E. Strom and K.M. Strom 'Surface Brightness and Colour Distributions of E and SO Galaxies II' Astron. J. Vol. 83, No. 7, pp. 733-763 (1978)
(3) S.A. Gregory and L.A. Thompson 'The Coma/A1367 Supercluster and its Environs' Ap. J. 222, pp. 784-799 (1978)
(4) G. Gavazzi Astron. and Astrophys. 69 (1978)

Abell 1377

(1) J.G. Godwin et al. 'Studies of Rich Clusters of Galaxies V' University of Oxford Preprint (submitted to Mon. Not. R.A.S.) 1979
(2) A.A. Hoffman and P. Crane 'A Photometric Study of Clusters of Galaxies' Ap. J. No. 215, pp. 379-400 (1977)

Abell 1656

(1) J.G. Godwin and J.V. Peach 'Studies of Rich Clusters of Galaxies - IV. Photometry of the Coma Cluster' Mon. Not. R.A.S., 181, pp. 323-337 (1977)
(2) G. Chincarini ed. H.J. Rood 'Size of the Coma Cluster' Nature Vol. 257 September 25, 1975
(3) S.A. Gregory and L.A. Thompson 'The Coma/A 1367 Supercluster' Ap. J. 222, pp. 784-799 (1978)
(4) G. Chincarini and H.J. Rood 'Space Distribution of Sc Galaxies: Clues to the Large Scale Structure' Ap. J. 230, pp. 648-654 (1979)
(5) L.A. Thompson 'The Angular Momentum Properties of Galaxies in Rich Clusters' Ap. J. 209, pp. 22-34 (1976)
(6) E.A. Valentin et al. 'A Westerbork Survey of Clusters of Galaxies IV. Observations of the Coma Cluster at 610 MHz' Astron. Astrophys. Suppl 28, pp. 333-349 (1977)

(7) Y.N. Pariiskii 'Detection of Hot Gas in the Coma Cluster of Galaxies'
Soviet Astronomy, Vol. 16, No. 6, (1973)

Abell 2065

(1) L. Macdonald 'Deep Sky Objects in the 24 inch Reflector'
Sky and Telescope January 1973
(2) J.G. Godwin et al. 'Studies of Rich Clusters of Galaxies V'
University of Oxford Preprint (submitted to Man. Not. R.A.S.) 1979
(3) R. Burnham Jnr. 'Burnham's Celestial Handbook Vol. 2'
Dover Inc., New York, 1978, p. 715

Abell 2151

(1) G.R. Burbridge and E.M. Burbridge 'The Hercules Cluster of Galaxies'
Ap. J. No. 130, pp. 629-640 (1959)
(2) B.A. Cooke et al. 'The X-Ray Emission from the Hercules Supercluster'
Astron. Astrophys. 58, pp L17-L19 (1977)
(3) J.G. Godwin et al. 'Studies of Rich Clusters of Galaxies V'
University of Oxford preprint (submitted to Mon. Not. R.A.S.) 1979
(4) H.C. Corwin Jnr. 'Notes on the Hercules Galaxy Cluster'
Pub. Astron. Soc. Pacific Vol. 83, pp. 320-326 (1971)

Abell 2197

(1) D. Carter 'The Optical Extent of Giant E and cD Galaxies'
Mon. Not. R.A.S., No. 178, pp. 137-148 (1977)
(2) L.A. Thompson 'The Angular Momentum Properties of Galaxies in Rich Clusters'
Ap. J. No. 209, pp. 22-34 (1976)

Abell 2199

(1) K.M. Strom and S.M. Strom 'Surface Brightness and Colour Distributions of E and SO Galaxies III'
Astron. J. Vol. 83, No. 11, pp. 1293-1330 (1978)
(2) D. Carter 'The Optical Extent of Giant E and cD Galaxies'
Mon. Not. R.A.S., No. 178, pp. 137-148 (1977)
(3) P. Gorenstein and W. Tucker 'Rich Clusters of Galaxies'
Scientific American May 1978

(3) Groups of Galaxies

NGC.383 Group

(1) S.F. Burch 'Multifrequency Radio Observations of 3C31: A large

Radio Galaxy with Jets and Peculiar Spectra'
Mon. Not. R.A.S. 181, pp. 599-610 (1977)
(2) C. Moss and R.J. Dickens 'Redshifts of Galaxies in the Cluster A262 and in the Region of the Pisces Group (NGC.383)'
Mon. Not. R.A.S. 178, pp. 701-715 (1977)

NGC.507 Group

(1) W.G. Tifft and K.A. Husman 'The NGC.507 Cluster of Galaxies'
Ap. J. 199, pp. 16-18 (1975)
(2) M.J. Thomson 'Observations of a Compact Group of Galaxies in Andromeda'
Webb Society Quarterly Journal No. 33, pp. 8-11 (1978)

NGC.5416 Group

(1) L.A. Thompson et al. 'The NGC.5416 Cluster of Galaxies'
Pub. Astron. Soc. Pacific Vol. 90, No. 538, pp. 644-649 (1978)

NGC.7320 Group

(1) G.B. Field, H.C. Arp and J.N. Bahcall 'The Redshift Controversy'
W.A. Benjamin Inc., Reading Mass. 1973

Bibliography: Catalogues and Atlases

G.O. Abell, 'The Distribution of Rich Clusters of Galaxies', Ap. J. Suppl. No. 3, pp. 211-288, University of Chicago Press, Chicago, 1958

H.C. Arp, 'Atlas of Peculiar Galaxies', California Institute of Technology, Pasadena, 1978

A. Becvar, 'Atlas of the Heavens-II, Catalog 1950.0 (Atlas Coeli Skalnate Pleso-II)', Sky Publishing Corp., Cambridge, Massachusetts, 1964

G. de Vaucouleurs and A. de Vaucouleurs, 'Reference Catalog of Bright Galaxies', University of Texas Press, Austin, 1964

G. de Vaucouleurs, A. de Vaucouleurs, and H.G. Corwin, Jr., 'Second Reference Catalogue of Bright Galaxies', University of Texas Press, Austin, 1976

J.L.E. Dreyer, 'Index Catalogue of Nebulae found in the years 1888-1894 with notes and corrections to the New General Catalogue', Mem. R.A.S., 51, 1895. 'Second Index Catalogue of Nebulae and Clusters of Stars found in the years 1895-1907 with notes and corrections to the New General Catalogue and to the Index Catalogue for 1888-1894', Mem R.A.S., 59, 1910. (IC 1 and IC 2 combined and reprinted by R.A.S., London, 1955)

J.L.E. Dreyer, 'New General Catalogue of Nebulae and Clusters of Stars', Mem. R.A.S., 49, 1953

A. Hirshfeld and R.W. Sinnott, 'Sky Catalogue 2000.0, Vol. 2, Deep-Sky Objects', Sky Publishing Corp., Cambridge, Massachusetts, forthcoming

J.H. Mallas and E. Kreimer, 'The Messier Album', Sky Publishing Corp., Cambridge, Massachusetts and Cambridge University Press, New York and Cambridge, England, 1978

P. Nilson, 'Uppsala General Catalogue of Galaxies', University of Uppsala, Uppsala, Sweden, 1973

A. Sandage, 'The Hubble Atlas of Galaxies', Carnegie Institution of Washington, Washington, D.C., 1961

A. Sandage and G.A. Tamman, 'A Revised Shapley--Ames Catalog of Bright Galaxies', Carnegie Institution of Washington, Washington, D.C.

G.E. Satterthwaite, P. Moore, and R.G. Inglis, eds., 'Norton's Star Atlas and Reference Handbook (Seventeenth edition)', Sky Publishing Corp., Cambridge, Massachusetts, 1978

C. Scovil, 'The AAVSO Variable Star Atlas', Sky Publishing Corp., Cambridge, Massachusetts, 1980

H. Shapley and H. Davies, 'Messier's Catalogue of Nebulae and Clusters', The Observatory 41, No. 529, 318

'Smithsonian Astrophysical Observatory Star Atlas of Reference Stars and Non-Stellar Objects', MIT Press, Cambridge, Massachusetts

'Smithsonian Astrophysical Observatory Star Catalog', Smithsonian Institution, Washington, D.C., 1966

J.W. Sulentic and W.G. Tifft, 'The Revised New General Catalogue of Non-Stellar Astronomical Objects', University of Arizona Press, Tucson, 1973

W. Tirion, 'Sky Atlas 2000.0', Sky Publishing Corp., Cambridge, Massachusetts and Cambridge University Press, Cambridge, England, 1981

H. Vehrenberg, 'Atlas of Deep-Sky Splendors', Treugesell Verlag, Dusseldorf West Germany, 1978 (To be published in a fourth edition by Sky Publishing Corp., Cambridge, Massachusetts)

B.A. Vorontsor-Velyaminov and V.P. Arhipova, 'Morphological Catalogue of Galaxies (4 volumes)', University of Moscow, Moscow, 1964

F. Zwicky, F.E. Herzog, and P. Wild, 'Catalogue of Galaxies and Clusters of Galaxies', California Institute of Technology, Pasadena, 1960-1968

Bibliography: General

M. Berry, 'Principles of Cosmology and Gravitation', Cambridge University Press, England, 1976

R.J. Dickens and J.E. Perry, eds., 'The Galaxy and the Local Group (Royal Greenwich Observatory Bulletin No. 182)', Herstmonceux, Royal Greenwich Observatory, 1976

T. Ferris, 'Galaxies', Thames and Hudson, London, and Sierra Club, San Francisco, 1980

K. Glyn Jones, 'Messier's Nebulae and Star Clusters', Faber & Faber, London, 1968

W.J. Kaufmann III, 'Galaxies and Quasars', W.H. Freeman and Co., San Francisco, 1979

M.S. Longair and J. Einasto, 'The Large Scale Structure of the Universe IAU Symposium No. 79)', D. Reidel Publishing Co., Boston, 1978

S. Mitton, 'Exploring the Galaxies', Faber & Faber, London, 1976 and Scribner, New York, 1981

A. Sandage, M. Sandage, and J. Kristian, eds., 'Galaxies and the Universe (Vol. 9 of Stars and Stellar Systems), University of Chicago Press, Chicago, 1975

J. Shakescraft, ed., 'The Formation and Dynamics of Galaxies (IAU Symposium No. 58)', D. Reidel Publishing Co., Boston, 1974

R.J. Tayler, 'Galaxies: Structure and Evolution', Crane, Russak and Co., New York, 1978

S. Van den Bergh, 'Clusters of Galaxies', Vistas in Astronomy Vol. 21, pp. 71-92, Pergamon Press, Elmsford, New York and Oxford, England